Jira 8 Essentials
Fifth Edition

Effective issue management and project tracking with the latest Jira features

Patrick Li

BIRMINGHAM - MUMBAI

Jira 8 Essentials
Fifth Edition

Commissioning Editor: Richa Tripathi
Acquisition Editor: Chaitanya Nair
Content Development Editor: Rohit Singh
Technical Editor: Royce John
Copy Editor: Safis Editing
Project Coordinator: Vaidehi Sawant
Proofreader: Safis Editing
Indexer: Priyanka Dhadke
Graphics: Alishon Mendonsa
Production Coordinator: Nilesh Mohite

First published: May 2011
Second edition: April 2013
Third edition: April 2015
Fourth edition: November 2016
Fifth edition: February 2019

Production reference: 1270219

Published by Packt Publishing Ltd.
Livery Place
35 Livery Street
Birmingham
B3 2PB, UK.

ISBN 978-1-78980-281-8

www.packtpub.com

`mapt.io`

Mapt is an online digital library that gives you full access to over 5,000 books and videos, as well as industry leading tools to help you plan your personal development and advance your career. For more information, please visit our website.

Why subscribe?

- Spend less time learning and more time coding with practical eBooks and Videos from over 4,000 industry professionals

- Improve your learning with Skill Plans built especially for you

- Get a free eBook or video every month

- Mapt is fully searchable

- Copy and paste, print, and bookmark content

Packt.com

Did you know that Packt offers eBook versions of every book published, with PDF and ePub files available? You can upgrade to the eBook version at `www.packt.com` and as a print book customer, you are entitled to a discount on the eBook copy. Get in touch with us at `customercare@packtpub.com` for more details.

At `www.packt.com`, you can also read a collection of free technical articles, sign up for a range of free newsletters, and receive exclusive discounts and offers on Packt books and eBooks.

Contributors

About the author

Patrick Li is the cofounder of AppFusions and works as a senior engineer there. AppFusions is one of the leading Atlassian experts, specializing in integration solutions with many enterprise applications and platforms, including IBM Connections, Jive, Google Apps, and more. He has worked in the Atlassian ecosystem for over 9 years, developing products and solutions for the Atlassian platform and providing expert consulting services. He has authored many books and video courses covering Jira 4 to 7. He has extensive experience in designing and deploying Atlassian solutions from the ground up and customizing existing deployments for clients across verticals such as healthcare, software engineering, financial services, and government agencies.

About the reviewer

Miroslav Kralik has over 16 years of experience in the IT field and has specialized knowledge in DevOps. He is a result-oriented and detail-oriented person with a passion for doing things in an easier and more effective way. He has also been part of various projects in which he trains divisions to automate IT processes and improve collaboration. He enjoys helping his customers to find solutions to their problems and creates best practices for them.

Packt is searching for authors like you

If you're interested in becoming an author for Packt, please visit authors.packtpub.com and apply today. We have worked with thousands of developers and tech professionals, just like you, to help them share their insight with the global tech community. You can make a general application, apply for a specific hot topic that we are recruiting an author for, or submit your own idea.

Table of Contents

Preface

Over the years, Jira has grown from a simple bug-tracking system designed for engineers to manage their projects to an all-purpose issue-tracking solution. As it has matured over time, Jira has become more than an application—it has transformed into a platform with a suite of other products that are built on top of it, enabling it to adapt and deliver value to a wide variety of use cases.

The term Jira now refers to a suite of products, including Jira Software, Jira Service Desk, and Jira Core. With this change, each product is more focused on what it does and the value it provides. It is now easier than ever for customers to choose the product best suited to their needs, whether they are running an Agile software development project, a customer support portal, or simply a generic task management system.

In this book, we will cover all the basics of Jira and the core capabilities of each product in the family, along with the add-ons that add additional features to the JIRA platform. Packed with real-life examples and step-by-step instructions, this book will help you become a Jira expert.

Who this book is for

This book will be especially useful for project managers, but it's also intended for other Jira users, including developers, and those in any other industry besides software development who would like to leverage Jira's powerful task management and workflow features to better manage their business processes.

What this book covers

Chapter 1, *Getting Started with Jira*, serves as an overall introduction to Jira by going over its high-level architecture. We will cover both fresh new deployments and how to upgrade from an existing deployment. This will also serve as the starting point of the project that readers will go through.

Chapter 2, *Using Jira for Business Projects*, covers using Jira for projects that are not based on software development, for example, a generic task management solution. This chapter focuses on use the basic features of Jira, which are offered through the Jira Core product, which is bundled with Jira Software.

Chapter 3, *Using Jira for Agile Projects*, covers features that are specific to Jira Software. This chapter focuses on using JIRA for software development projects, especially using Agile methodologies such as Scrum and Kanban.

Chapter 4, *Issue Management*, introduces issues, which are the cornerstone of using Jira. The focus is to make sure users understand issues and what they do. You will also learn how to make each of the features available and customize them further beyond the out-of-box settings.

Chapter 5, *Field Management*, introduces fields, and specifically how to use custom fields to customize Jira for more effective data collection. You will learn how to create different types of custom fields and their usages, and how to control field behaviors such as visibility and rendering options.

Chapter 6, *Screen Management*, introduces screens. You will learn how to create new screens from scratch and specify which fields (system and custom) will be displayed. We will also cover the complex scheme mappings to apply new screens to projects.

Chapter 7, *Workflow and Business Process*, explores the most powerful feature offered by Jira, workflows. The concept of issue life cycles is introduced, and various aspects of workflows are explained. This chapter also explores the relationship between workflows and other various Jira aspects that have been previously covered, such as screens. The concept of Jira add-ons is also briefly touched upon in the sample project, using some popular add-ons.

Chapter 8, *Emails and Notifications*, talks about emails and how Jira can use it to send notifications with end users. We will start by explaining how Jira sends out notifications to users, and then how Jira can process incoming emails to create, comment, and also update issues.

Chapter 9, *Securing Jira*, explains Jira's security model, starting with how to manage users, groups, and roles. Readers will then learn Jira's security hierarchy of how permissions are managed. Lastly, we will look at integrating JIRA with LDAP, a common requirement with most enterprise organizations.

`Chapter 10`, *Searching, Reporting, and Analysis*, focuses on doing more with data collected by Jira, including searching, reporting, and using dashboards. Readers will also learn how to make this data and reports available outside of Jira, either via email, or by displaying them in other applications.

`Chapter 11`, *Jira Service Desk*, introduces one of the new add-ons, called Jira Service Desk, which allows you to run Jira as a customer support portal. Readers will learn how to use Jira Service Desk to run and manage a support queue internally while at the same time communicating effectively with customers with the add-on.

To get the most out of this book

The installation package used in this book is the Windows Installer standalone distribution, which you can get directly from Atlassian at `https://www.atlassian.com/software/jira/download` for JIRA Software and `https://www.atlassian.com/software/jira/service-desk/`download for JIRA Service Desk.

You will also need additional software, including the Java SDK, which you can get from `http://www.oracle.com/technetwork/java/javase/downloads/index.html`, and MySQL, which you can get from `http://dev.mysql.com/downloads`.

Download the example code files

You can download the example code files for this book from your account at `www.packt.com`. If you purchased this book elsewhere, you can visit `www.packt.com/support` and register to have the files emailed directly to you.

You can download the code files by following these steps:

1. Log in or register at `www.packt.com`.
2. Select the **SUPPORT** tab.
3. Click on **Code Downloads & Errata**.
4. Enter the name of the book in the **Search** box and follow the onscreen instructions.

Once the file is downloaded, please make sure that you unzip or extract the folder using the latest version of:

- WinRAR/7-Zip for Windows
- Zipeg/iZip/UnRarX for Mac
- 7-Zip/PeaZip for Linux

We also have other code bundles from our rich catalog of books and videos available at `https://github.com/PacktPublishing/`. Check them out!

Download the color images

We also provide a PDF file that has color images of the screenshots/diagrams used in this book. You can download it here: `https://www.packtpub.com/sites/default/files/downloads/9781789802818_ColorImages.pdf`.

Conventions used

There are a number of text conventions used throughout this book.

`CodeInText`: Indicates code words in text, database table names, folder names, filenames, file extensions, pathnames, dummy URLs, user input, and Twitter handles. Here is an example: "You should see both the `New Employee` and `Termination` issue types."

A block of code is set as follows:

```
<Resource name="mail/JiraMailServer"
  auth="Container"
  type="javax.mail.Session"
```

When we wish to draw your attention to a particular part of a code block, the relevant lines or items are set in bold:

```
<security-constraint>
  <web-resource-collection>
    <web-resource-name>all-except-attachments</web-resource-name>
    <url-pattern>*.js</url-pattern>
    <url-pattern>*.jsp</url-pattern>
```

Any command-line input or output is written as follows:

```
create database jiradb character set utf8;
```

Bold: Indicates a new term, an important word, or words that you see onscreen. For example, words in menus or dialog boxes appear in the text like this. Here is an example: "There is a **Create another** option beside the **Create** button."

Warnings or important notes appear like this.

 Tips and tricks appear like this.

Get in touch

Feedback from our readers is always welcome.

General feedback: If you have questions about any aspect of this book, mention the book title in the subject of your message and email us at customercare@packtpub.com.

Errata: Although we have taken every care to ensure the accuracy of our content, mistakes do happen. If you have found a mistake in this book, we would be grateful if you would report this to us. Please visit www.packt.com/submit-errata, selecting your book, clicking on the Errata Submission Form link, and entering the details.

Piracy: If you come across any illegal copies of our works in any form on the Internet, we would be grateful if you would provide us with the location address or website name. Please contact us at copyright@packt.com with a link to the material.

If you are interested in becoming an author: If there is a topic that you have expertise in and you are interested in either writing or contributing to a book, please visit authors.packtpub.com.

Reviews

Please leave a review. Once you have read and used this book, why not leave a review on the site that you purchased it from? Potential readers can then see and use your unbiased opinion to make purchase decisions, we at Packt can understand what you think about our products, and our authors can see your feedback on their book. Thank you!

For more information about Packt, please visit packt.com.

Section 1: Introduction to Jira 8

In this section, you will learn how to set up a new Jira 8 instance from scratch, followed by how to use Jira for your business and agile projects.

The following chapters will be covered in this section:

- Chapter 1, *Getting Started with Jira*
- Chapter 2, *Using Jira for Business Projects*
- Chapter 3, *Using Jira for Agile Projects*

Getting Started with Jira

1

In this chapter, we will start with a high-level view of Jira, going through each of the components that make up the overall application. We will then examine the various deployment options, system requirements for Jira 8, and the platforms/software that are supported. Finally, we will get our hands dirty by installing our very own Jira 8 from scratch with the installation wizard. Finally, we will also cover some post-installation steps, such as setting up SSL to secure our new instance.

By the end of this chapter, you will have learned about the following:

- The different offerings from the Jira product family
- The overall architecture of Jira
- The basic hardware and software requirements to deploy and run Jira
- Platforms and applications supported by Jira
- Installing Jira and all of the required software
- Post-installation configuration options to customize your Jira

Jira Core, Jira Software, and Jira Service Desk

Starting with Jira 7, Jira is split into three different products, and the term Jira now refers to the common platform that all these products are built on. The three products that make up the Jira family are as follows:

- **Jira Core**: This is similar to the classic Jira (also known as JIRA), with all the field customizations and workflow capabilities. This is perfect for general-purpose task management.

- **Jira Software**: This is Jira Core with agile capabilities (previously known as JIRA Agile). This is well-suited for software development teams that want to use agile methodologies, such as Scrum and Kanban.
- **Jira Service Desk**: This is Jira Core with service desk capabilities. This is designed for running Jira as a support ticketing system, with a simplified user interface for the end users, and a focus on customer satisfaction with SLA goals.

As you can see, Jira Core is at the heart, providing all the base functionalities, such as user interface customization, workflows, and email notifications, while Jira Software and Jira Service Desk add specialized features on top of it.

In this book, we will mostly focus on Jira Software. However, since Jira Core provides many of the common features, most of the knowledge is also applicable to Jira Core, and features that are only available to Jira Software will be highlighted. For this reason, the term *Jira* will be used to cover both Jira Core and Jira Software, unless a distinction is required. We will also cover Jira Service Desk in `Chapter 11`, *Jira Service Desk*.

The Jira architecture

Installing Jira is simple and straightforward. However, it is important for you to understand the components that make up the overall architecture of Jira and the installation options that are available. This will help you make an informed decision and be better prepared for future maintenance and troubleshooting, as well as establishing some common terminologies that are often used by the user community and Atlassian support representatives.

High-level architecture

Atlassian provides a comprehensive overview of the Jira architecture at `https://developer.atlassian.com/server/jira/platform/architecture-overview`. However, with regards to the day-to-day administration and utilization of Jira, we do not need to go into the details of this; the information provided can be overwhelming at first glance. For this reason, we have summarized a high-level overview, which highlights the most important components in the architecture, as shown in the following diagram:

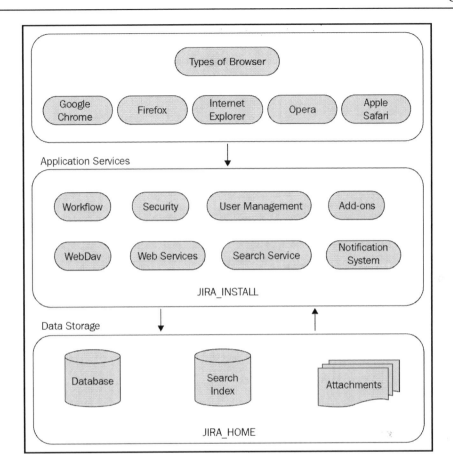

Web browsers

Jira is a web application, so there is no need for users to install anything on their machines. All they need is a web browser that is compatible with Jira. The following table summarizes the browser requirements for Jira:

Browsers	Compatibility
Internet Explorer	11 and Edge
Mozilla Firefox	Latest stable versions
Safari	Latest stable versions on mac OSX
Google Chrome	Latest stable versions
Mobile	Mobile Safari on iOS only Mobile Chrome

Application services

The application services layer contains all the functions and services provided by Jira. These services include various business functions, such as workflow and notification, which will be discussed in depth in Chapter 7, *Workflow and Business Process*, and Chapter 8, *Emails and Notifications*, respectively. Other services, such as REST/Web Service, provide integration points to other applications, and the OSGi service provides the base add-on framework to extend Jira's functionalities.

Data storage

The data storage layer stores persistent data in several places within Jira. Most business data, such as projects and issues, are stored in a relational database. Content such as uploaded attachments and search indexes are stored in the filesystem in the JIRA_HOME directory, which we will talk about in the next section. The underlying relational database that's used is transparent to users, and you can migrate from one database to another with ease, as referenced at https://confluence.atlassian.com/adminjiraserver/switching-databases-938846867.html.

The Jira installation directory

The Jira installation directory is where you install Jira. It contains all the executable and configuration files of the application. Jira neither modifies the contents of the files in this directory during runtime, nor does it store any data files inside the directory. The directory is used primarily for execution. For the remainder of this book, we will refer to this directory as JIRA_INSTALL.

The Jira home directory

The Jira home directory contains key data and configuration files specific to each Jira instance, such as Jira's database connectivity details. As we will see later in this chapter, setting the path to this directory is part of the installation process.

There is a one-to-one relationship between Jira and this directory. This means that each Jira instance must have only one home directory, and each directory can serve only one Jira instance. In the old days, this directory was sometimes called the data directory. It has now been standardized as the Jira Home. It is for this reason that, for the remainder of this book, we will refer to this directory as JIRA_HOME.

The JIRA_HOME directory can be created anywhere on your system, or even on a shared drive, but it cannot be a sub-directory of JIRA_INSTALL. It is recommended to use a fast disk drive with low network latency to get the best performance from Jira.

This separation of data and application makes tasks such as maintenance and future upgrades an easier process. Within JIRA_HOME, there are several subdirectories that contain vital data, as shown in the following table:

Directory	Description
data	This directory contains data that is not stored in the database, for example, uploaded attachment files.
export	This directory contains the automated backup archives created by Jira. This is different from a manual export executed by a user; manual exports require the user to specify where to store the archive.
import	This directory contains the backups that can be imported. Jira will only load backup files from this directory.
log	This directory contains Jira log files, which are useful for tracking down errors. Some of the key log files include the following: • atlassian-jira.log: Information about Jira Software and the Jira Core application • atlassian-servicedesk.log: Information about the Jira Service Desk application • atlassian-jira-security.log: Information about user sessions, logins, and logouts
plugins	This directory is where installed plugins (also known as add-ons) are stored. Add-ons will be discussed further in later chapters.
caches	This directory contains cache data that Jira uses to improve its performance at runtime. For example, search indexes are stored in this directory.
tmp	This directory contains temporary files created at runtime, such as file uploads.

When Jira is running, the JIRA_HOME directory is locked. When Jira shuts down, it is unlocked. This locking mechanism prevents multiple Jira instances from reading/writing to the same JIRA_HOME directory and causing data corruption.

Jira locks the JIRA_HOME directory by writing a temporary file called jira-home.lock into the root of the directory. During shutdown, this file will be removed. Occasionally, however, Jira may fail to remove this file, such as during an ungraceful shutdown. In this case, you can manually remove this locked file to unlock the directory so that you can start up Jira again.

You can manually remove the locked file to unlock the JIRA_HOME directory if Jira fails to clean it up during the shutdown.

System requirements

Just like any other software application, a set of base requirements needs to be met before you can install and run Jira. Therefore, it is important for you to be familiar with these requirements so that you can plan out your deployment successfully. Note that these requirements are for a behind-the-firewall deployment, also known as the **Jira Server** or **Jira Data Center.** The main difference between the two is that Jira Data Center allows for clustering, so you can have additional benefits such as high availability and better scalability. Atlassian also offers a cloud-based alternative called **Jira Cloud**, available at `https://www.atlassian.com/software#cloud-products`.

Hardware requirements

For evaluation purposes, where there will only be a small number of users, Jira will run happily on any server that has a 1.5 GHz processor and 1 GB to 2 GB of RAM. As your Jira usage grows, a typical server will have a quad core 2 GHz+ CPU and 4 GB of RAM dedicated to the Jira application, and at least 10 GB of hard disk space for your database.

For production deployment, as in most applications, it is recommended that you run Jira on its own dedicated server. There are many factors that you should consider when deciding the extent of the resources to allocate to Jira; keep in mind how Jira will scale and grow. When deciding on your hardware needs, you should consider the following:

- The number of active (concurrent) users in the system
- The number of issues and projects in the system
- The number of configuration items, such as custom fields and workflows
- The number of concurrent users, especially during peak hours

It can be difficult at times to estimate these figures. As a reference, a server running with over 2.0 quad core CPU and 4 GB of RAM will be sufficient for most instances with around 200 active users. If you start to get into thousands of active users, you will need to have at least 8 GB of RAM allocated to Jira (JVM). Once you have gone beyond a million of issues and thousands of active users for a single Jira instance, simply adding raw system resources (vertical scaling) will start yield diminishing returns. In such cases, it is often better to consider using the data center edition of Jira, which offers better scalability by allowing you to have multiple instances clustered together (horizontal scaling), with the added benefit of providing high availability.

Officially, Jira only supports x86 hardware and 64-bit derivatives of it. When running Jira on a 64-bit system, you will be able to allocate more than 4 GB of memory to Jira, which is the limit if you are using a 32-bit system. If you are planning to deploy a large instance, it is recommended that you use a 64-bit system.

Software requirements

Jira has four requirements when it comes to software. It needs a supported operating system and a Java environment. It also needs an application server to host and serve its contents and a database to store all of its data. In the following sections, we will discuss each of these requirements and the options that you have to install and run Jira. You can find the latest information online at `https://confluence.atlassian.com/ adminjiraserver/supported-platforms-938846830.html`.

Operating systems

Jira supports most of the major operating systems, so the choice of which operating system to run Jira on becomes a matter of expertise, comfort, and, in most cases, the existing organization's IT infrastructure and requirements.

The operating systems supported by Atlassian are Windows and Linux. There is a Jira distribution for mac OSX, but this is mostly for evaluation purposes only. Cloud-based deployment options are also available for **Amazon Web Services** (**AWS**) and Microsoft Azure. However, with these cloud options, there are restrictions for components such as database support.

With both Windows and Linux, Atlassian provides an executable installer wizard package, which bundles all the necessary components to simplify the installation process (only available for standalone distribution). There are minimal differences when it comes to installing, configuring, and maintaining Jira on different operating systems. If you do not have any preferences and would like to keep initial costs down, CentOS Linux is a good choice.

Java platforms

Jira is a Java-based web application, so it needs to have a Java environment installed. This can be a **Java Development Kit (JDK)** or a **Java Runtime Environment (JRE)**. The executable installer that comes with Windows or Linux contains the necessary files and will install and configure the JRE for you. However, if you want to use archive distributions, you will need to make sure that you have the required Java environment installed and configured.

Jira 8 requires Java 8 (also known as 1.8). If you run Jira on an unsupported Java version, including its patch version, you may run into unexpected errors. The following table shows the supported Java versions for JIRA:

Java platforms	Support status
Oracle JDK/JRE	Java 8 (1.8)

With the recent licensing changes made to Oracle JDK by Oracle, efforts are currently underway to add support for OpenJDK. However, at the time of writing, OpenJDK is not officially supported.

Databases

Jira stores all its data in a relational database. While you can run Jira with **H2 Database**, the in-memory database that comes bundled with Jira, it is prone to data corruption. You should only use this to set up a new instance quickly for evaluation purposes, where no important data will be stored. For this reason, it is important that you use a proper database such as MySQL for production systems.

Most relational databases available on the market today are supported by Jira, and there are no differences when you install and configure Jira. Just like operating systems, your choice of database will come down to your IT staff's expertise, experience, and established corporate IT standards. If you run Windows as your operating system, then you probably want to go with the Microsoft SQL Server. On the other hand, if you run Linux, then you should consider Oracle (if you already have a license), MySQL, or PostgreSQL.

The following table summarizes the databases that are currently supported by Jira. It is worth mentioning that both MySQL and PostgreSQL are open source products, so they are excellent options if you are looking to minimize your initial investments.

Database	Support status
MySQL	MySQL 5.6 and newer. Note that neither MariaDB nor PerconaDB are supported. This requires the latest JDBC driver.
PostgreSQL	PostgreSQL 9.4 and newer. This requires the latest PostgreSQL JDBC (9.4) driver.
Microsoft SQL Server	SQL Server 2012 and newer. This requires the latest Microsoft JDBC (6.2.1) driver.
Oracle	Oracle 12c R1. This requires the latest Oracle driver.
H2	This is bundled with the standalone distribution, for evaluation purposes only.

Take special note of the driver requirement on each database, as some drivers that come bundled with the database vendor or Jira itself (for example, PostgreSQL) are not supported and will need to be replaced with the appropriate versions.

Application servers

Jira 8 officially only supports Apache Tomcat as the application server. While it is possible to deploy Jira into other JEE compliant application servers, you will be doing this at your own risk, and it is not recommended.

The following table shows the versions of Tomcat supported by Jira 8:

Application server	Support status
Apache Tomcat	Tomcat 8.5.32 and newer. You should not deploy other applications or multiple Jira instances into the same Tomcat server.

Installation options

Jira comes in two flavors—an executable installer and a `TAR.GZ` or `ZIP` archive. The executable installer provides a wizard-driven interface that will walk you through the entire installation process. It even comes with a Java installer to save you some time. The archive flavor contains everything except for a Java installer, which means you will have to install Java yourself. You will also need to perform some post-installation steps manually, such as configure Jira as a service. However, you do get the advantage of learning what really goes on under the hood.

Installing and configuring Jira

Now that you have a good understanding of the overall architecture of Jira, the basic system requirements, and the various installation options, we are ready to deploy our own Jira instances.

In the following exercise, we will be installing and configuring a fresh Jira instance for a small production team. We will perform our installation on a Windows platform with a MySQL database server. If you are planning to use a different platform or database, refer to the vendor documentation on installing the required software for your platform.

In this exercise, you will do the following:

- Install a fresh instance of Jira Software
- Connect Jira to a MySQL database

We will continue to use this Jira instance in subsequent chapters and exercises as we build our help desk implementation.

For our deployment, we will use the following:

- Jira Software server distribution 8
- MySQL 5.7.13
- Microsoft Windows 7

Installing Java

Since we will be using the installer package that's bundled with Java, you can skip this section. However, if you are using the archive option, you need to make sure that you have Java installed on your system.

Jira 8 requires JRE version 8 (1.8) to run. You can verify the version of Java you have by running the following command in a Command Prompt:

```
java -version
```

The preceding command tells us which version of Java is running on your system, as shown in the following screenshot:

If you do not see a similar output, then chances are you do not have Java installed. You will need to perform the following steps to set up your Java environment. We will start by installing JDK on your system:

1. Download the latest JDK from `http://www.oracle.com/technetwork/java/javase/downloads/index.html`.

 At the time of writing, the latest version of Java 8 is JDK 8 Update 192.

2. Double-click on the downloaded installation file to start the installation wizard.
3. Select where you would like to install Java, or you can simply accept the default values. The location where you install JDK will be referred to as JAVA_HOME for the remainder of this book.

4. Create a new environment variable named JAVA_HOME with the value set to the full path of the location where you installed Java. You can do this as follows:

> 1. Open the System Properties window by holding down your Windows key and pressing the *Pause* key on your keyboard.
>
> 2. Select the **Advanced system settings** option.
>
> 3. Click on the **Environment Variable** button from the new popup:

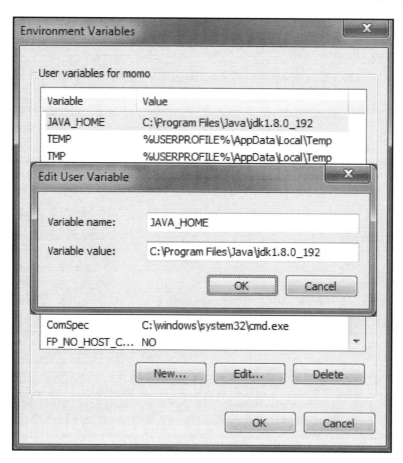

5. Edit the PATH environment variable and append the following to the end of its current value:

```
;%JAVA_HOME%\bin
```

6. Test the installation by typing the following command in a new Command Prompt:

```
java -version
```

This will display the version of Java installed, provided everything is done correctly. In Windows, you have to start a new Command Prompt after you have added the environment variable to see the change.

Installing MySQL

The next step is to prepare an enterprise database for your Jira installation. Jira requires a fresh database. If, during the installation process, Jira detects that the target database already contains data, it will not proceed. If you already have a database system installed, then you may skip this section.

To install MySQL, simply perform the following steps:

1. Download MySQL from `http://dev.mysql.com/downloads`, select **MySQL Community Server**, and then select the **MSI installer** for Windows.

 At the time of writing, the latest version of MySQL is 5.7.13.

2. Double-click on the downloaded installation file to start the installation wizard.
3. Click on **Install MySQL Products** on the welcome screen.
4. Read and accept the license agreement and click on the **Next** button.
5. Select the **Server only** option on the next screen. If you are an experienced database administrator, you can choose to customize your installation. Otherwise, just accept the default values for all subsequent screens.
6. Configure the MySQL root user password. The username will be `root`. Do not lose this password, as we will be using it in the next section.
7. Complete the configuration wizard by accepting the default values.

Preparing MySQL for Jira

Now that you have MySQL installed, you need to create a user for Jira to connect MySQL with as you should never use the default root user, and then create a fresh database for Jira to store all its data:

1. Start the MySQL Command Line Client by navigating to **Start | All Programs | MySQL | MySQL Server 5.7 | MySQL 5.7 Command Line Client**.
2. Enter the MySQL root user password you set during installation.
3. Use the following command to create a database:

   ```
   create database jiradb character set utf8;
   ```

4. Here, we are creating a database called `jiradb`. You can name the database anything you like. As you will see later in this chapter, this name will be referenced when you connect JIRA to MySQL. We have also set the database to use `utf8` character encoding, as this is a requirement for JIRA. Using the following command, you need to ensure that the database uses the InnoDB storage engine to avoid data corruption:

   ```
   grant all on jiradb.* to 'jirauser'@'localhost'
   identified by 'jirauserpassword';
   ```

   We are doing several things here. First, we create a user called `jirauser` and assign the password `jirauserpassword` to them. You should change the username and password to something else.

   We have also granted all the privileges to the user for the `jiradb` database that we just created so that the user can perform database operations, such as create/drop tables and insert/delete data. If you have named your database something other than `jiradb`, then make sure that you change the command so that it uses the name of your database.

   This allows you to control the fact that only authorized users (specified in the preceding command) are able to access the Jira database to ensure data security and integrity.

5. To verify your setup, exit the current interactive session by issuing the following command:

   ```
   quit;
   ```

6. Start a new interactive session with your newly created user:

```
mysql -u jirauser -p
```

7. You will be prompted for a password, which you set up in the preceding command as jirauser.

8. Use the following command:

```
show databases;
```

   This will list all the databases that are currently accessible by the logged-in user. You should see jiradb among the list of databases.

9. Examine the jiradb database by issuing the following commands:

```
use jiradb;
show tables;
```

The first command connects you to the jiradb database, so all of your subsequent commands will be executed against the correct database.

The second command lists all the tables that exist in the jiradb database. Right now, the list should be empty, since no tables have been created for JIRA; but don't worry, as soon as we connect to Jira, all the tables will automatically be created.

Installing Jira

With the Java environment and database prepared, you can now move on to installing Jira. Normally, there are only two steps:

1. Download and install the Jira application
2. Run through the Jira setup wizard

Obtaining and installing Jira

The first step is to download the latest stable release of Jira. You can download Atlassian Jira from http://www.atlassian.com/software/jira/download.

The Atlassian website will detect the operating system you are using and automatically suggest an installation package for you to download. If you intend to install Jira on a different operating system from the one you are currently on, make sure that you select the correct operating system package.

As we mentioned earlier, with Windows, there is a Windows installer package and a self-extracting archive package. For the purpose of this exercise, we will use the installer package (Windows 64-bit Installer):

1. Double-click on the downloaded installation file to start the installation wizard and click on the **Next** button to continue:

2. Select the **Custom Install (recommended for advanced users)** option and click on the **Next** button to continue. Using the custom installation will let us decide where to install Jira and will also provide numerous configuration options:

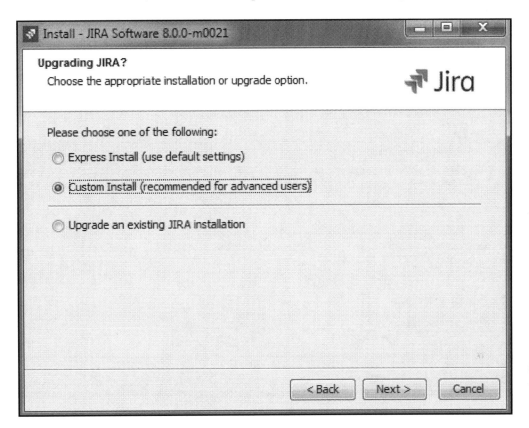

3. Select the directory where Jira will be installed. This will become the `JIRA_INSTALL` directory. Click on the **Next** button to continue:

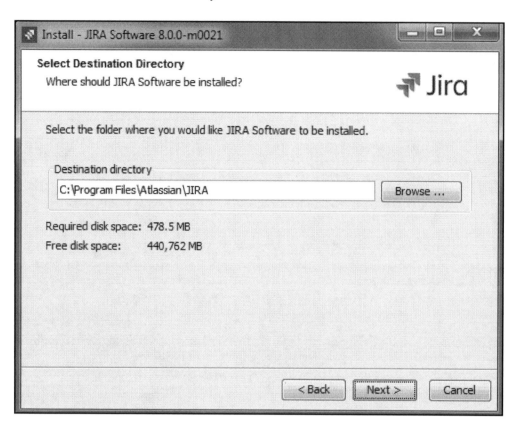

4. Select where Jira will store its data files, such as attachments and log files. This will become the JIRA_HOME directory. Click on the **Next** button to continue:

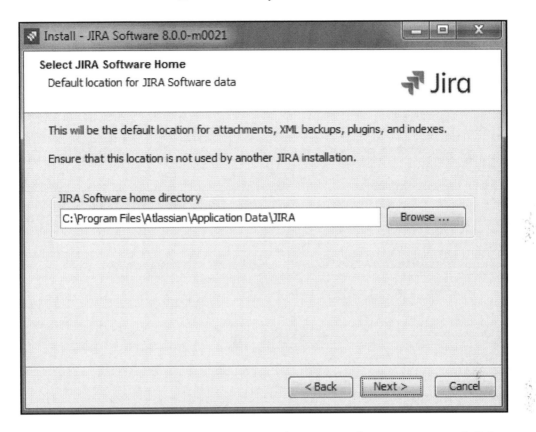

5. Select where you would like to create shortcuts to the start menu and click on the **Next** button to continue.

6. In the **Configure TCP Ports** step, we need to select the port on which Jira will be listening for incoming connections. By default, Jira will run on port 8080. If 8080 has already been taken by another application, or if you want Jira to run on a different port such as port 80, select the **Set custom value for HTTP and Control ports** option and specify the port numbers you want to use. Click on the **Next** button to continue:

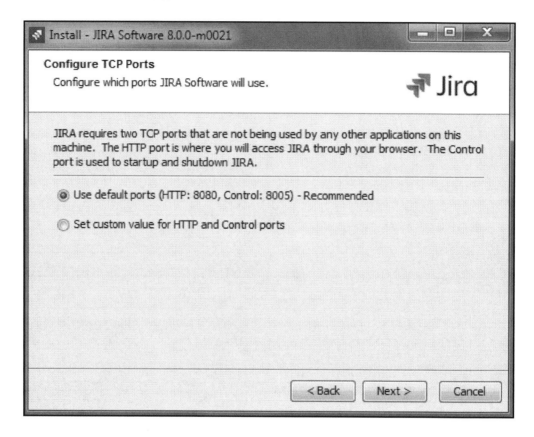

7. Select whether you would like Jira to run as a service. If you enable this option, Jira will be installed as a system service and can be configured to start automatically with the server; refer to the *Starting and stopping Jira* section for more details:

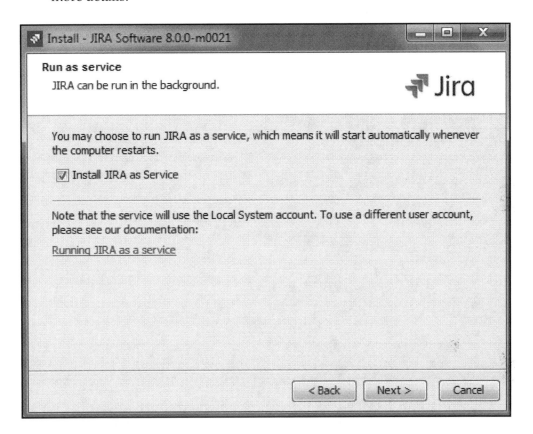

8. For the final step, review all the installation options and click on the **Install** button to start the installation:

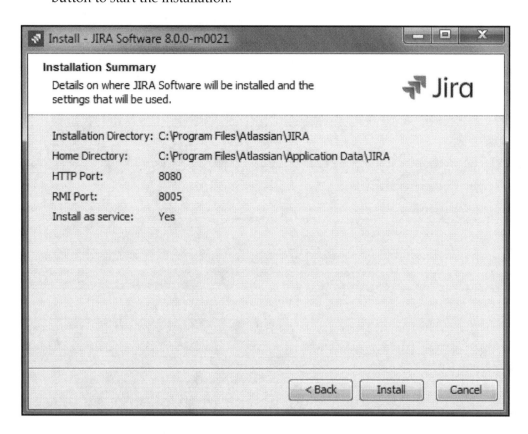

9. Once the installation is complete, check the **Launch Jira Software in browser** option and click on **Finish**. This will close the installation wizard and open up your web browser to access Jira. This might take a few minutes to load as Jira starts up for the first time:

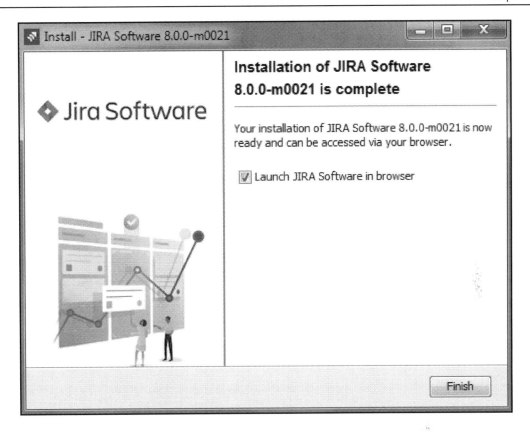

Installing MySQL driver

Jira 8 comes bundled with the MySQL database driver, so you can skip this section. However, if you do need to manually install the driver for some reason, such as the driver file got corrupted or accidentally deleted, you can download the required driver from http://dev.mysql.com/downloads/connector/j/. Once downloaded, you can install the driver by copying the driver JAR file into the JIRA_INSTALL/lib directory. After that, you need to restart Jira. If you have installed Jira as a Windows service in step 9, refer to the *Starting and stopping Jira* section.

 Make sure that you select the **Platform Independent** option and download the JAR or TAR archive.

The Jira setup wizard

Jira comes with an easy-to-use setup wizard that will walk you through the installation and configuration process in six simple steps. You will be able to configure the database connections, default language, and much more. You can access the wizard by opening `http://localhost:<port number>` in your browser, where the `<port number>` is the number you have assigned to Jira in step six of the installation process.

The steps are explained in the following sections.

Step one

In the first step of the wizard, we need to select how we want Jira to be set up. Since we are installing Jira for production use, we will select the **I'll set it up myself** option, as demonstrated in the following screenshot:

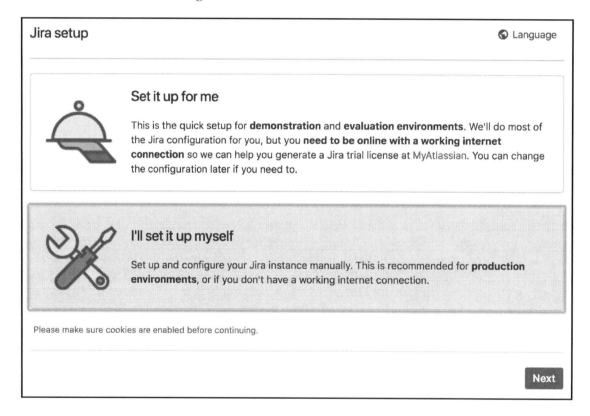

Step two

For the second step, we will need to select the database we want to use. This is where we configure Jira to use the MySQL database we created earlier in this chapter. If you select the **Built In** option, Jira will use its bundled in-memory H2 database, which is good for evaluation purposes. If you want to use a proper database, such as in our case, you should select the **My Own Database** option:

Database setup

Database Connection	○ Built In (for evaluation or demonstration)
	● My Own Database (recommended for production environments)
	Built in database can be migrated to a database of your own later.
	Learn more about connecting Jira to a database.
Database Type	MySQL 5.7+ ▾
Hostname	
	Hostname or IP address of the database server.
Port	3306
	TCP Port Number for the database server.
Database	
	The name of the database to connect to.
Username	
	The username used to access the database.
Password	
	The password used to access the database.

Next   Test Connection

The **Built In** option is great for getting Jira up and running quickly for evaluation purposes.

After you have selected the **My Own Database** option, the wizard will expand for you to provide the database connection details. If you do not have the necessary database driver installed, Jira will prompt you for it.

Once you have filled in the details for your database, it's a good idea to first click on the **Test Connection** button to verify that Jira is able to connect to the database. If everything is set up correctly, Jira will report a success message. You should be able to move onto the next step by clicking on the **Next** button. This may take a few minutes, as Jira will now create all the necessary database objects. Once this is done, you will be taken to step three of the wizard.

Step three

In the third step, you will need to provide some basic details about this Jira instance. Once you have filled in the requisite fields, click on **Next** to move on to step four of the wizard:

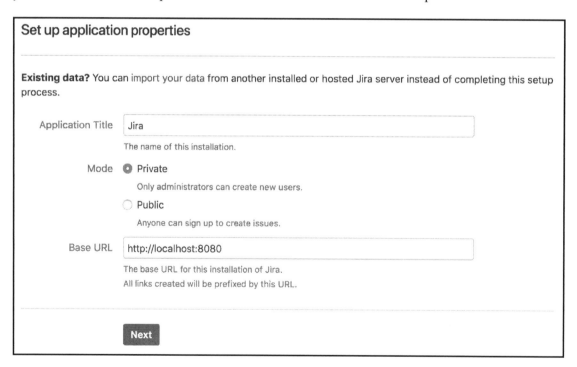

Step four

In the fourth step, we need to provide a license key for Jira. If you have already obtained a license from Atlassian, you can paste it into the **Your License Key** text box. If you do not have a license, you can generate an evaluation license by clicking on the **generate a Jira trial license** link. The evaluation license will grant you access to Jira's full set of features for one month. After the evaluation period ends, you will lose the ability to create new issues, but you can still access your data:

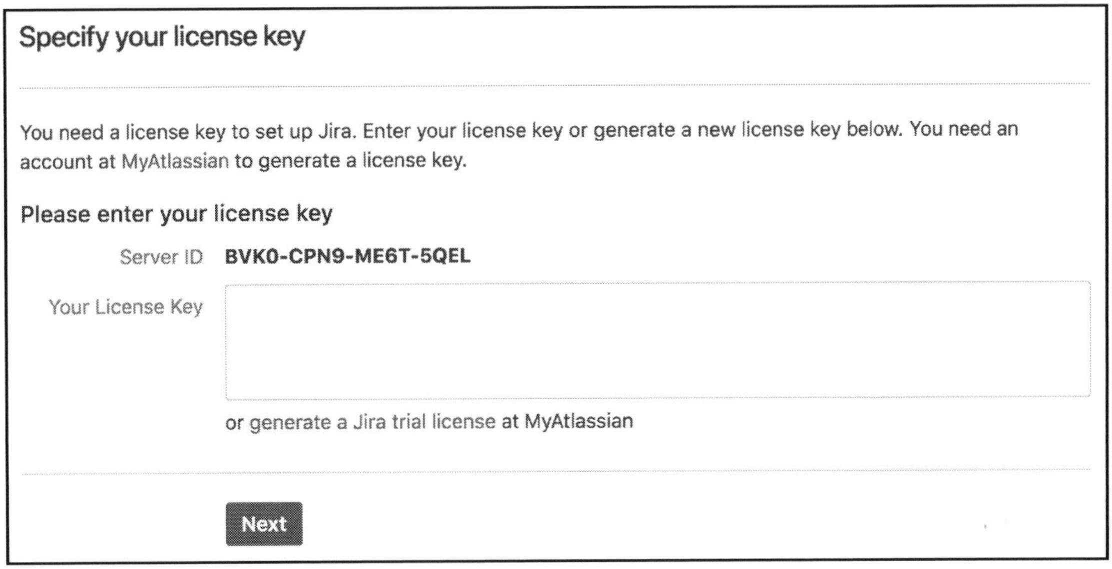

Specify your license key

You need a license key to set up Jira. Enter your license key or generate a new license key below. You need an account at MyAtlassian to generate a license key.

Please enter your license key

Server ID   **BVK0-CPN9-ME6T-5QEL**

Your License Key

or generate a Jira trial license at MyAtlassian

Next

Step five

In the fifth step, you will be setting up the administrator account for Jira. It is important that you keep the account details somewhere safe and not lose the password. Since Jira only stores the hashed value of the password instead of the actual password itself, you will not be able to retrieve it. Fill in the administrator account details and click on **Next** to move on to the sixth step:

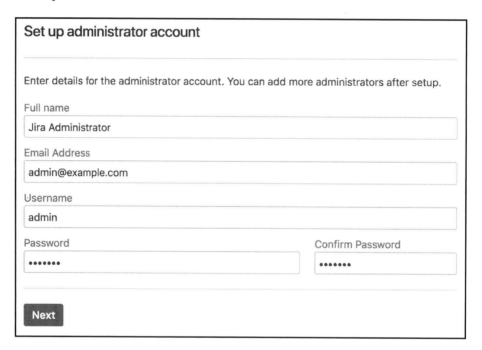

 This account is important and it can help you troubleshoot and fix problems later on. Do not lose it!

Step six

In the sixth step, you can set up your email server details. Jira will use the information configured here to send out notification emails. Notification is a very powerful feature in Jira and one of the primary methods by means of which Jira communicates with users. If you do not have your email server information handy, you can skip this step for now by selecting the **Later** option and clicking on **Finish**. You can configure your email server settings later, as you will see in Chapter 8, *Emails and Notifications*:

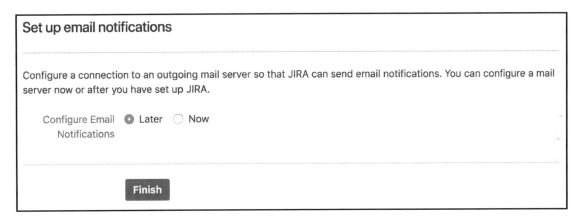

Congratulations! You have successfully completed your Jira setup. You should now see the welcome page, and be automatically logged in as the administrator user you created in step five. On the welcome page, you will need to set up a few user preferences, such as the default language and profile picture. Follow the onscreen prompts to set up the account, and once you are done, you should be presented with the options to create a sample project, a new project from scratch, or import project data from other sources, as shown in the following screenshot:

Starting and stopping Jira

Since we used the Windows Installer, Jira is installed as a Windows service. Therefore, you can start, stop, and restart it via the Windows services console by navigating to **Start** | **Control Panel** | **Administrative Tools** | **Services**. In the services console, look for Atlassian JIRA, and you will be able to stop and start the application, as shown in the following screenshot:

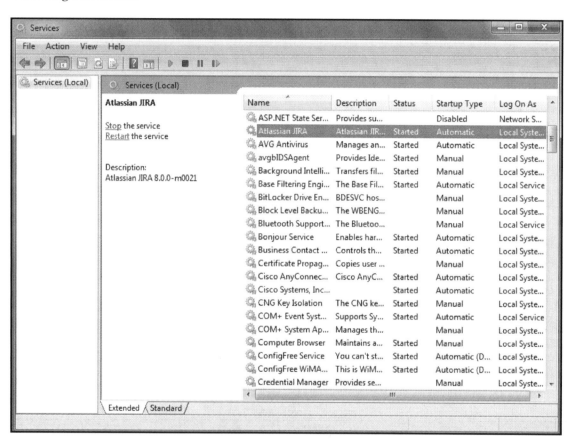

Post-installation configurations

The post-installation configuration steps are optional, depending on your needs and environment. If you set up Jira for evaluation purposes, you probably do not need to perform any of the following steps, but it is always good practice to be familiar with these as a reference.

 You will need to restart Jira after making the changes that will be discussed in the next section.

Increasing Jira's memory

The default memory setting for Jira is usually sufficient for a small- to medium-sized deployment. As Jira's adoption rate increases, you will find that the amount of memory allocated by default is no longer enough. If Jira is running as a Windows service, as we described in this chapter, you can increase the memory as follows:

1. Find the JIRA Windows service name. You can do this by going to the Windows services console and double-clicking on the **Atlassian JIRA** service. The service name will be the part after **//RS//** in the **Path to executable** field, for example, `JIRA150215215627`.

2. Open a new Command Prompt and change the current working directory to the `JIRA_INSTALL/bin` directory.

3. Run the following command by substituting the actual service name for Jira:

```
tomcat7w //ES//<JIRA Windows service name>
```

4. Select the **Java** tab, update the **Initial memory pool** and **Maximum memory pool** sizes, and click on **OK**:

5. Restart Jira to apply the change.

If you are not running Jira as a Windows service, you need to open the `setenv.bat` file (for Windows) or the `setenv.sh` (for Linux) file in the `JIRA_INSTALL/bin` directory. Then, locate the following lines:

```
set JVM_MINIMUM_MEMORY="384m"
set JVM_MAXIMUM_MEMORY="768m"
```

Change the value for the two parameters and restart Jira. Normally, 4 GB (4,096 m) of memory is enough to support a fairly large instance of Jira used by hundreds of users.

Make sure that you have sufficient physical RAM available before allocating instances to Jira.

Changing Jira's port number and context path

As part of the installation process, the installation wizard prompted us to decide which port JIRA should listen to for incoming connections. If you have accepted the default value, it is port 8080. You can change the port setting by locating and opening the `server.xml` file in a text editor in the `JIRA_INSTALL/conf` directory. Let's examine the relevant contents of this file:

```
<Server port="8005" shutdown="SHUTDOWN">
```

This line specifies the port for the command to shutdown Jira/Tomcat. By default, it is port 8005. If you already have an application that is running on that port (usually another Tomcat instance), you need to change this to a different port:

```
<Connector port="8080" protocol="HTTP/1.1">
```

This line specifies which port Jira/Tomcat will be running on. By default, it is port 8080. If you already have an application that is running on that port, or if the port is unavailable for some reason, you need to change it to another available port:

```
<Context path="/jira" docBase="${catalina.home}/atlassian-jira"
reloadable="false" useHttpOnly="true">
```

This line allows you to specify the context that Jira will be running under. By default, the value is empty, which means JIRA will be accessible from `http://hostname:portnumber`. If you decide to specify a context, the URL will be `http://hostname:portnumber/context`. In our example here, Jira will be accessible from `http://localhost:8080/jira`.

Configuring HTTPS

By default, Jira runs with a standard, non-encrypted HTTP protocol. This is acceptable if you are running Jira in a secured environment, such as an internal network. However, if you plan to open up access to Jira over the internet, you will need to tighten up security by encrypting sensitive data, such as usernames and passwords that are being sent, by enabling HTTPS (HTTP over SSL).

For a standalone installation, you will need to perform the following tasks:

1. Obtain and install a certificate
2. Enable HTTPS on your application server (Tomcat)
3. Redirect traffic to HTTPS

First, you need to get a digital certificate. This can be obtained from a certification authority, such as VeriSign (CA certificate), or a self-signed certificate that's been generated by you. A CA certificate will not only encrypt data for you, but also identify your copy of Jira to the users. A self-signed certificate is useful when you do not have a valid CA certificate and you are only interested in setting up HTTPS for encryption. Since a self-signed certificate is not signed by a certification authority, it is unable to identify your site to the public and users will be prompted with a warning that the site is untrusted when they first visit it. However, for evaluation purposes, a self-signed certificate will suffice until you can get a proper CA certificate.

For the purpose of this exercise, we will create a self-signed certificate to illustrate the complete process. If you have a CA certificate, you can skip the following steps.

Java comes with a handy tool for certificate management, called `keytool`, which can be found in the `JIRA_HOME\jre\bin` directory if you are using the installer package. If you are using your own Java installation, then you can find it in `JAVA_HOME\bin`.

To generate a self-signed certificate, run the following commands from a Command Prompt:

```
keytool -genkey -alias tomcat -keyalg RSA
keytool -export -alias tomcat -file file.cer
```

This will create a keystore (if one does not already exist) and export the self-signed certificate (`file.cer`). When you run the first command, you will be asked to set the password for the keystore and Tomcat. You need to use the same password for both. The default password is `changeit`. You can specify a different password of your choice, but then you have to let Jira/Tomcat know, as we will see later.

Now that you have your certificate ready, you need to import it into your trust store for Tomcat to use. Again, you will use the `keytool` application in Java:

```
keytool -import -alias tomcat -file file.cer
JIRA_HOME\jre\lib\security\cacerts
```

This will import the certificate into your trust store, which can be used by JIRA/Tomcat to set up HTTPS.

To enable HTTPS on Tomcat, open the `server.xml` file in a text editor from the `JIRA_INSTALL/conf` directory. Locate the following configuration snippet:

```
<Connector port="8443" maxHttpHeaderSize="8192" SSLEnabled="true"
maxThreads="150" minSpareThreads="25" maxSpareThreads="75"
enableLookups="false" disableUploadTimeout="true"
acceptCount="100" scheme="https" secure="true"
clientAuth="false" sslProtocol="TLS"      useBodyEncodingForURI="true"/>
```

This enables HTTPS for Jira/Tomcat on port 8443. If you have selected a different password for your keystore, you will have to add the following line to the end of the preceding snippet before the closing tag:

```
keystorePass="<password value>"
```

The last step is to set up Jira so that it automatically redirects from a non-HTTP request to an HTTPS request. Find and open the `web.xml` file in the `JIRA_INSTALL/atlassian-jira/WEB-INF` directory. Then, add the following snippet to the end of the file before the closing `</web-app>` tag:

```
<security-constraint> <web-resource-collection> <web-resource-name>all-
except-attachments</web-resource-name> <url-pattern>*.js</url-pattern>
<url-pattern>*.jsp</url-pattern>
    <url-pattern>*.jspa</url-pattern>
    <url-pattern>*.css</url-pattern>
    <url-pattern>/browse/*</url-pattern>
  </web-resource-collection>
  <user-data-constraint>
    <transport-guarantee>CONFIDENTIAL</transport-guarantee>
  </user-data-constraint>
</security-constraint>
```

Now, when you access Jira with a normal HTTP URL, such as `http://localhost:8080/jira`, you will be automatically redirected to its HTTPS equivalent, `https://localhost:8443/jira`.

Summary

Jira is a powerful, yet simple, application, as reflected in its straightforward installation procedures. You have a wide variety of options available so that you can choose how you would like to install and configure your copy. You can mix-and-match different aspects, such as operating systems and databases, to best suit your requirements. The best part is that you can have a setup that consists entirely of open source software, which will bring down the cost and provide you with a reliable infrastructure at the same time.

In the next chapter, we will start to explore various aspects of Jira. The following chapters, starting with projects, will talk about key concepts in any Jira installation.

Using Jira for Business Projects 2

Jira initially started off as a bug-tracking system, helping software development teams to better track and manage the problems/issues in their projects. As the product evolved, people started using Jira for other purposes; some use it as a general-purpose, task-management solution, others use it as a customer support portal, and some financial institutions even use Jira to track their portfolios. So, starting with Jira 7, Atlassian has made many improvements to help make Jira ubiquitous by introducing three distinct solutions—Jira Core, Jira Software, and Jira Service Desk.

In this chapter, we will take a look at projects and project types, focusing on the most basic Jira project type—business. We will then take a look at the various user interfaces that Jira has for working with projects, both as an administrator and an everyday user. We will also introduce permissions for the first time in the context of projects and will expand on this in later chapters. Business being the most basic project type, most of the concepts covered in this chapter will be applicable to the more specialized types.

By the end of this chapter, you will have learned the following:

- Jira project types and templates
- Different user interfaces for project management in Jira
- How to create new projects in Jira
- How to import data from other systems into Jira
- How to manage and configure a project
- How to manage components and versions

Understanding project types

Project types define the features available for your projects as well as the user interface that will be used to present information within the projects. Each project type also comes with one or more templates, with a set of predefined configurations to help you get started quickly. The following screenshot shows you the project types and their templates from an out-of-the-box Jira Software installation:

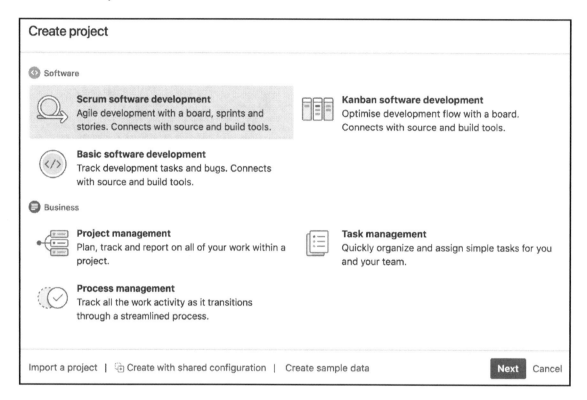

The **Software** and **Business** templates are included in Jira Software. If you are running Jira Core, you will only have templates under **Business**.

If you create a project using the **Scrum software development** template under the **Software** project type, your project will come with a Scrum board and a set of configurations designed to work with the Scrum methodology. On the other hand, if you choose the **Task management** template under the **Business** project type, your project will have a different user interface designed for task management.

Business projects

As we have already seen, Jira comes with a number of project types, depending on what applications you have installed. The business project type, being part of Jira Core, is available to all installations.

Business projects are very similar to how Jira worked before Jira 7, when it worked as a highly customizable, generic task-tracking system. The focus of the business project type and its templates is to enable users to easily create tasks and track and report their progress.

Out of the box, you have three built-in templates, each with a set of predefined configurations such as workflows and fields, to give you some idea of how to set up your projects. You can use them as is or customize them further, based on your needs.

Jira permissions

Before we start working with projects in Jira, we need to first understand a little bit about permissions. Permissions are a big topic, and we will cover them in detail in Chapter 9, *Securing Jira*. For now, we will briefly talk about permissions related to creating and deleting, administering, and browsing projects.

In Jira, users with the Jira administrator global permission will be able to create and delete projects. By default, users in the Jira administrators group have this permission, so the administrator user we created during the installation process in Chapter 1, *Getting Started with Jira*, will be able to create new projects. We will refer to this user and any other users with this permission as a Jira administrator.

For any given project, users with the **Administer Project** permission for that project will be able to administer the project's configuration settings. As we will see in the later sections of this chapter, this means that users with this permission will have access to the **Project Administration** interface for a given project. This allows them to update the project's details and configurations. By default, the Jira administrator will have this permission.

If a user needs to browse the contents of a given project, then they must have the **Browse Project** global permission for that project. This means that the user will have access to the **Project Browser** interface for the project. By default, the Jira administrator will have this permission.

Creating projects

The easiest way to create a new project is to select the **Create Project** menu option from the **Projects** drop-down menu from the top navigation bar. This will bring up the **Create project** dialog. Note that, as we explained, you need to be a Jira administrator (such as the user we created during installation) to create projects. This option is only available if you have the permission.

From the **Create project** dialog, select the template you want to use under the **Business** heading and click on **Next**. On the next page, Jira will display predefined configurations for the template you have selected. In our example, we have selected the **Task management** template, so Jira provides us with two issue types and a very simple workflow with two steps. Click on the **Select** button to continue, as shown in the following screenshot:

 Jira will use the selected template to create new sets of configurations called schemes for our new project.

 You can change these configurations once the project is created.

For the third and final step, we need to provide the new project's details. Jira will help you validate the details, for example, by making sure that the project key is unique and conforms to the required format. After filling in the project details, click on the **Submit** button to create the new project, as show in the following screenshot:

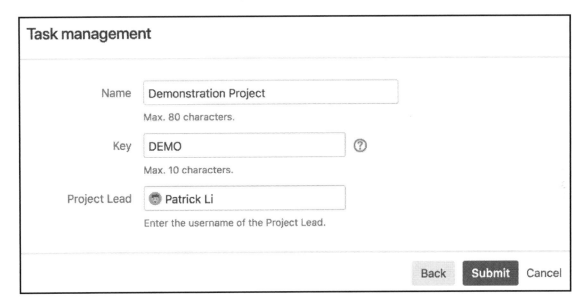

The following table lists the information you need to provide when creating a new project:

Field	Description
Name	This is a unique name for the project.
Key	This is a unique identity key for the project. As you type the name of your project, Jira will auto-fill the key based on the name, but you can replace the auto-generated key with one of your own. You will be able to change the key later.The project key will also become the first part of the issue key for issues created in the project.
Project Lead	The lead of the project can be used to auto-assign issues. Each project can have only one lead. This option is available only if you have more than one user in Jira.

Once you have created the new project, you will be taken to the **Project Browser** interface, which we will discuss in the forthcoming sections.

You may have noticed that in the **Create project** dialog, there are the following three additional options at the bottom:

- **Import a project**: This allows you to import data from another compatible issue tracking system or data export, such as Bugzilla, GitHub, or CSV file, into a new or existing Jira project. We will cover this later in the chapter.
- **Create with shared configuration**: This allows you to create a new project based on the configurations of an existing project. This is a great way for you to create a project based on a standard set of configurations quickly.
- **Create sample data**: This allows you to create a new project and populate it with some sample issues, so you can start exploring the various features offered by different project type templates.

Changing the project key format

When creating new projects, you may find that the project key needs to be in a specific format and length. By default, the project key needs to adhere to the following criteria:

- It should contain at least two characters
- It cannot be more than 10 characters in length
- It should contain only characters—that is, no numbers

You can change the default format to have less restrictive rules.

 These changes are for advanced users only.

First, to change the project key length, go through the following steps:

1. Browse to the Jira administration console
2. Select the **System** tab and then the **General configuration** option
3. Click on the **Edit Settings** button
4. Change the value for the **Maximum project key size** option to a value between 2 and 255 (inclusive) and click on the **Update** button to apply the change

Changing the project key format is a bit more complicated. Jira uses a regular expression to define what the format should be. To change the project key format, go through the following steps:

1. Browse to the Jira administration console
2. Select the **System** tab and then the **General configuration** option
3. Click on the **Advanced Settings** button
4. Hover over and click on the (**[A-Z][A-Z]+**) value for the `jira.projectkey.pattern` option. For example, if you want to use numbers in your project key, you can use (**[A-Z][A-Z0-9]+**)
5. Enter the new regular expression for the project key format and click on **Update**

There are a few rules when it comes to setting the project key format, as follows:

- The key must start with a letter
- All letters must be uppercase—that is, **[A-Z]**
- Only letters, numbers, and the underscore character can be used
- The new pattern must be compatible with all existing projects

Project user interfaces

There are two distinctive interfaces for projects in Jira. The first interface is designed for everyday users, providing useful information on how the project is going using reports, statistics, and agile boards. This interface is called the **project browser**.

The second interface is designed for project administrators to control project configuration settings, such as permissions and workflows, and is called **Project Administration**.

After creating a project, the first interface you see will be the **Project Browser**. We will start our discussion by looking at this interface and then move on to the **Project Administration** interface.

Project browser

The project browser is the interface that most users will use with Jira. It acts as the home page of the project, providing useful information (such as recent activities in the project) reports, and information from other connected systems, such as source control and continuous integration. The actual project browser interface depends on the project type, so it will vary from project to project. For example, the Scrum software development project will display the issue backlog or an agile board as its default view, as shown in the following screenshot:

For business projects such as general task management, the project browser will display an activity stream showing the latest updates in the project, as shown in the following screenshot:

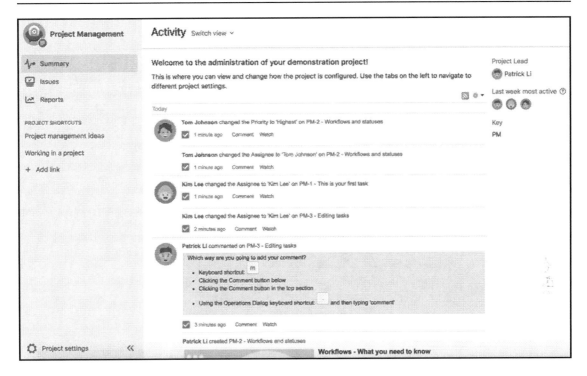

To access the **Project Browser** interface, simply select the project from the **Projects** drop-down or the project list via the **View All Projects** option. A business project's project browser contains the following tabs, listed here along with their descriptions:

Project Browser tab	Description
Summary	This displays a quick overview of the project. It comes in two views: an activity view and a statistics view.
Issues	This displays a breakdown of issues in the project grouped by attributes, such as priority and status.
Reports	This contains a number of built-in and custom reports that you can generate based on the issues in the project. The types of report available will vary depending on the project type.
Versions	This displays the summary of the versions of the project. This tab is only available when versions are configured.
Components	This displays a summary of components and their related issues. This tab is only available when components are configured for the project.
Project Shortcuts	This lets you add custom links that will act as bookmarks to provide more information on the project.

We will look at some of these tabs in greater detail in the next sections.

The Summary tab

The **Summary** tab provides you with a single-page view of the project you are working on. For business projects, it provides an activity view, which will display the latest activities that are happening in the project, and a statistics view, which provides you with a number of useful breakdowns of the issues within the project. For example, **Unresolved: By Assignee** lets you know how many open issues are assigned to each user, allowing the project team to better plan their resource allocation, as shown in the following screenshot:

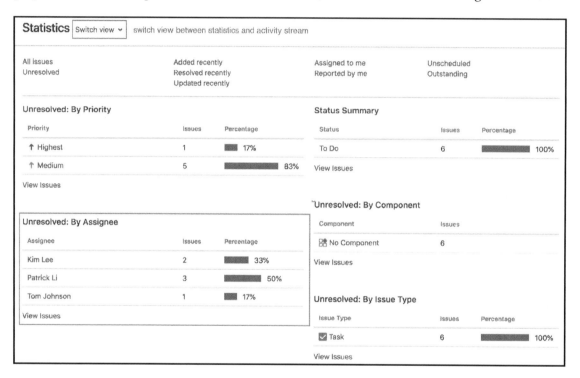

The Issues tab

The **Issues** tab, by default, lists all open issues in the project. It also contains a number of other predefined filters you can use to look for issues. From the list, you can select an individual issue and get more information, as shown in the following screenshot:

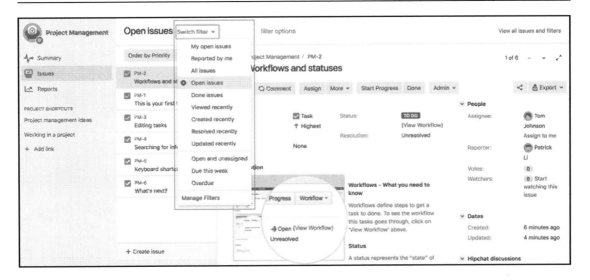

The Versions and Components tabs

The **Versions** and **Components** tabs list all the available versions and components that have been configured for this project, respectively. These two tabs are only visible if the project contains versions and/or components.

The Project Administration interface

The **Project Administration** interface is where project administrators can manage the settings and configurations of their projects. For example, you can change the project's name, select what issue types will be available for the project, and manage a list of components within the project. Only users with the **Administer Projects** permission for a given project will be able to access this interface.

Starting with Jira 7.3, project administrators are given more control over how their projects should be configured. Prior to Jira 7.3, configurations such as workflows and priorities were system-level configurations. As we will see in later chapters, Jira 8 continues with these changes to empower project administrators by removing the reliance on one or two Jira system administrators to manage all the configurations.

To access the **Project Administration** interface, go through the following steps:

1. Go to the project browser for the project you want to administer.
2. Select the **Project settings** option in the bottom-left corner. If you do not see the option, then you do not have the necessary permission.

From the **Project Administration** interface, you will be able to perform the following key operations:

- Update project details, such as the project name, description, avatar, and type
- Manage what users see when working on the project, such as issue types, fields, and screens
- Configure the workflows used by the project
- Control permission settings and notifications
- Manage the list of available components and versions

 If you are not a Jira administrator (that is, if you do not have the Jira administrator global permission), you can only view the current project's configuration. We will cover permissions in `Chapter 9`, *Securing Jira*.

The preceding key operations are shown in the following screenshot:

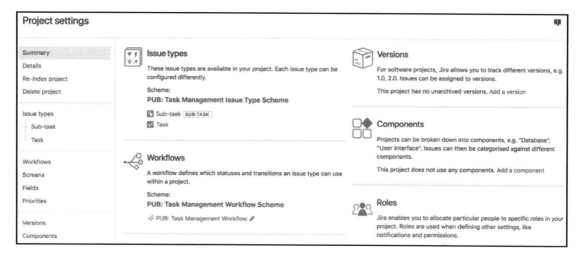

As a project administrator, this is where you will be applying customizations to your project. We will cover each of the customizations in later chapters.

The project details tab

The first group consists of a number of tabs, including **Summary**, **Details**, **Re-index project**, and **Delete project**, as described in the following list:

- **Summary**: Displays a single-page view of all the current configuration settings for the project.
- **Details**: Allows you to change the project's general information, including the project key, type, avatar, description, and category.
- **Re-index project**: Re-indexes the project to update the search index for issues in this project. Perform this when there are configuration changes made to the project, such as new fields are added.
- **Delete project**: Deletes the project and all its issues. This operation cannot be undone.

The Components tab

The **Components** tab is where project administrators can manage the components for their projects. Components can be thought of as subsections that make up the full project. In a business project, components can be various business functions or operations that need to be completed. As shown in the following screenshot, there are three components configured in the current project:

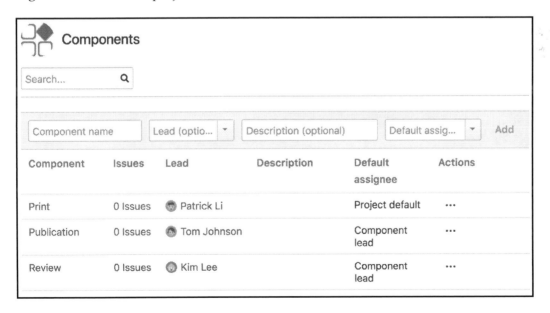

Components are project specific in Jira. This means that components from one project cannot be used in a different project. This also allows each project to maintain its own sets of components. A component has four pieces of information, as shown in the following table:

Field	Description
Name	This is a unique name for the component.
Description	This is an optional description to offer more explanation as to what the component is for.
Lead	This is an optional field where you can select a single user as the lead for the component. For example, in a software project, this can be the main developer for the component.
Default assignee	This tells Jira when an issue is created without the assignee being selected, and if the issue has a component, then Jira will auto-assign the issue to the selected default assignee.

Creating components

Unlike some older versions of Jira, you can create new components directly on the page in more recent Jira editions. You can do this by going through the following steps:

1. Browse to the **Components** tab for the project.
2. Enter a unique name for the component. Jira will let you know whether the name is already taken.
3. Enter a short description for the new component.
4. Select a user to be the lead of the component. Just start typing and Jira will prompt you with options that meet the criteria.
5. Select the default assignee option for the component.
6. Click on **Add** to create the new component.

Once you have created the new component, it will be added to the list of existing components. When a component is first created, it will be placed at the top of the list. If you refresh the page, the list will then be ordered alphabetically.

The component lead and default assignee

One of the useful features of components is the ability to assign a default assignee to each component. This means that when a user creates an issue with a component and sets the assignee as **Automatic**, Jira will be able to automatically assign the issue based on the component selected. This is a very powerful feature for organizations where members of various teams often do not know each other. Therefore, when it comes to assigning issues at creation time, they often find it difficult to decide who to assign them to. This feature can be set up so that the lead of the component becomes the default assignee and the issues raised can then be delegated to other members of the team.

> If the issue has more than one component with a default assignee, the assignee for the first component in alphabetical order will be used.

The Versions tab

Like the **Components** tab, the **Versions** tab allows project administrators to manage versions for their projects. Versions serve as milestones for a project. In project management, versions represent points in time. While versions may seem less relevant for projects that are not product-oriented, versions can still be valuable when it comes to managing and tracking the progress of issues and work.

As with components, versions also have a number of attributes, as shown in the following table:

Field	Description
Name	This is a unique name for the version.
Description	This is an optional description to offer more explanation as to what the version is for.
Start Date	This is a date indication that states when work on this version is expected to start.
Release Date	This is an optional date that will be marked as the scheduled date on which to release the version. Versions that are not released according to the release date will have the date highlighted in red.

Creating versions

Creating new versions is as simple as providing the necessary details for the new version and clicking on the **Add** button. Simply go through the following steps:

1. Browse to the **Versions** tab for the target project.
2. Enter a unique name for the version (for example, `1.1.0`, `v2.3`). Jira will let you know whether the name is already taken.
3. Enter a short description for the new version.
4. Select the date on which the version starts and will be released using the date picker.
5. Click on **Add** to create the new version.

Unlike components, versions will not be ordered automatically by Jira, so you will have to manually maintain the order. To change the order of versions all you have to do is hover your mouse pointer to the left of the version, and you will be able to drag the version up and down the list.

Managing versions

When you hover over a version, you will notice that there is a little cog icon to the right. If you click on that, you will have the option to do the following:

Option	Description
Release	This will mark the version as released, meaning that it is completed or shipped. When you release a version, Jira will automatically check to make sure that all the issues are completed for the selected version. If there are incomplete issues, you will be prompted to either ignore those issues or push them to a different version. If the version has already been released, it will change to **Unreleased**.
Build and Release	This is similar to the **Released** option, but it also performs a build through an integrated build system, such as Atlassian Bamboo, if there are any software codes. The version will only be released if the build is successful. This option is not available if the version is already released, or Bamboo is not integrated with Jira.
Archive	This will mark the version as archived, meaning that the version is stored away until further notice. When a version is archived, you cannot release or delete it until it is unarchived.
Delete	This will delete the version from Jira. Again, Jira will search for issues that are related to this version and ask you whether you would like to move these issues to a different version.
Edit	This will let you update the version's details such as its target release date.

The options described in the preceding table can be seen in the following screenshot:

 There is also a **Merge versions** feature that allows you to merge multiple versions together. Merging versions will move issues from one version to another.

Other tabs

There are a number of other tabs in the **Project Administration** interface. We will not explore these tabs in this chapter, as they will each be covered in their own chapters later. We will, however, take a look at what each tab does, as shown in the following table:

Tab	Description
Issue types	This controls the types of issues that users can create for the project. For example, this may include bugs, improvements, and tasks. **Issue types** will be covered in Chapter 4, *Issue Management*.
Workflows	This controls the workflow issues that we will go through. Workflows consist of a series of steps that usually mimic the existing processes that are in place for the organization. Workflows will be covered in Chapter 7, *Workflow and Business Process*.
Screens	Screens are what users see when they view, create, and edit issues in Jira. Screens will be covered in Chapter 6, *Screen Management*.

Fields	These are what Jira uses to capture data from users when they create issues. Jira comes with a set of default fields and the Jira administrator is able to add additional fields as needed. Fields will be covered in `Chapter 5`, *Field Management*.
Users and roles	Project administrators can define roles in the project and assign users to them. These roles can then be used to control permissions and notifications. Roles will be covered in `Chapter 9`, *Securing Jira*.
Permissions	As we have already seen, permissions define who can perform certain tasks or have access in Jira. Permissions will be covered in `Chapter 9`, *Securing Jira*.
Issue Security	Jira allows users to control who can view the issues they created by selecting the issue security level. Issue security will be covered in `Chapter 9`, *Securing Jira*.
Notifications	Jira has the ability to send out email notifications when certain events occur. For example, when an issue is updated, Jira can send out an email to alert all users who participate about the issue of the change. Notifications will be covered in `Chapter 8`, *Emails and Notifications*.

Importing data into Jira

Jira supports importing data directly from many popular issue-tracking systems and common data formats. All the importers have a wizard-driven interface, guiding you through a series of steps. These steps are mostly identical, but have a few differences. Generally speaking, there are four steps when importing data into Jira, as follows:

1. Select your source data. For example, if you are importing from CSV, it will select a CSV file. If you are importing from Bugzilla, it will provide Bugzilla database details.
2. Select a destination project where the imported issues will go. This can be an existing project or a new project created on the fly.
3. Map fields from the other source to Jira fields.
4. Map field values from the other source to Jira field values. This is usually required for select-based fields, such as the priority field, or select list custom fields.

Importing data through CSV

Jira comes with a CSV importer, which lets you import data in the **comma-separated value** (**CSV**) format. This is a useful tool if you want to import data from a system that is not directly supported by Jira, since most systems are able to export their data in CSV.

 We recommend that you do a trial import on a test instance first. You can find out more about CSV file format at `https://confluence.atlassian.com/adminjiracloud/importing-data-from-csv-776636762.html`.

Go through the following steps to import data through a CSV file:

1. Select the **Import External Project** option from the **Projects** drop-down menu.
2. Click on the **CSV importer** option. This will start the import wizard.
3. First, you need to select the CSV file that contains the data you want to import by clicking on the **Choose File** button.
4. After you have selected the source file, you can also expand the **Advanced** section to select the file encoding and delimiter used in the CSV file. There is also a **Use an existing configuration** option, which we will talk about this later in this section.

5. Click on the **Next** button to proceed, as shown in the following screenshot:

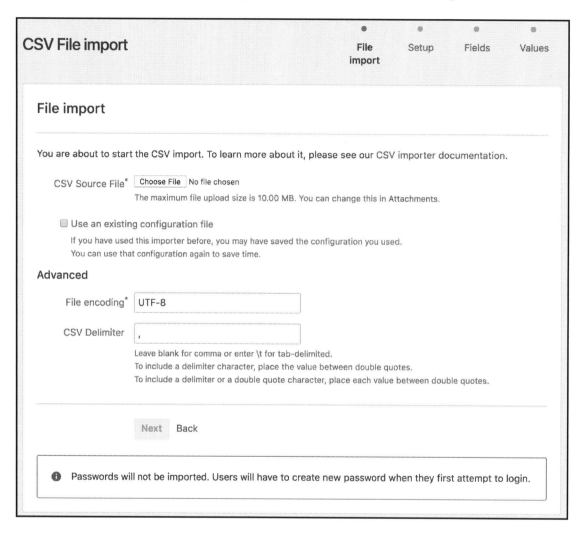

6. For the second step, you need to select the project you want to import our data into. You can also select the **Create New** option to create a new project on the fly.

7. If your CSV file contains date-related data, ensure that you enter the format used in the **Date format** field.
8. Click on the **Next** button to proceed, as shown in the following screenshot:

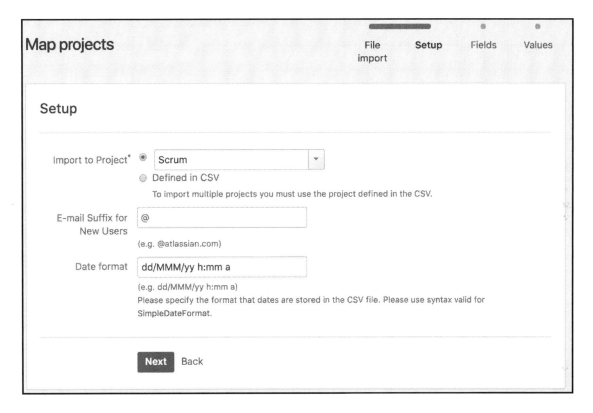

9. For the third step, you need to map the CSV fields to the fields in Jira. Not all fields need to be mapped. If you do not want to import a particular field, simply leave the corresponding Jira field selection as **Don't map this field**.
10. For fields that contain data that needs to be mapped manually, such as select list fields, you need to check the **Map field value** option. This will let you map the CSV field value to the Jira field value so they can be imported correctly. If you do not manually map these values, they will be copied over as is.

11. Click on the **Next** button to proceed, as shown in the following screenshot:

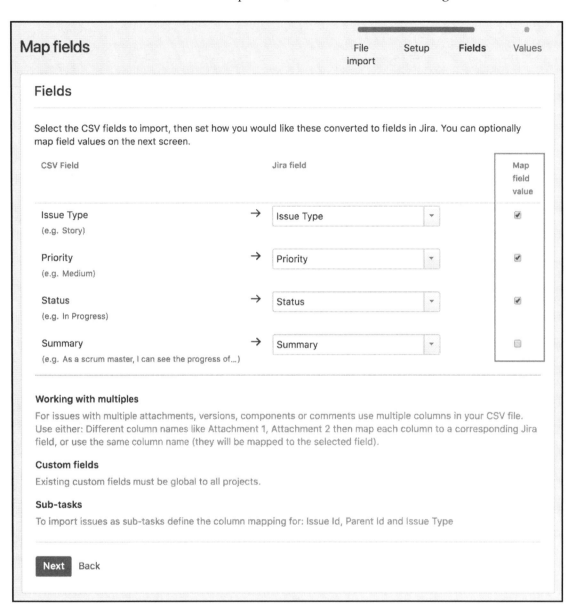

12. For the last step, you need to map the CSV field value to the Jira field value. This step is only required if you have checked the **Map field value** option for a field in step 10.

13. Enter the Jira field value for each CSV field value.
14. Once you are done with mapping field values, click on the **Begin Import** button to start the actual import process, as shown in the following screenshot. Depending on the size of your data, this may take some time to complete:

15. Once the import process completes, you will get a confirmation message that tells you the number of issues that have been imported, as shown in the following screenshot. This number should match the number of records you have in the CSV file:

 1 projects and 23 issues imported successfully!

What now?

You can download a detailed log of this import. You can also save the configuration for future use.

Import another project.

On the last confirmation screen, you can click on the **download a detailed log** link to download the full log file containing all the information for the import process. This is particularly useful if the import was not successful.

You can also click on the **save the configuration** link, which will generate a text file containing all the mappings you have done for this import. If you need to run a similar import in the future, you can even use this import file so that you will not need to manually remap everything again. To use this configuration file, check the **Use an existing configuration file** option in step 4.

As we can see, Jira's project importer makes importing data from other systems simple and straightforward. However, you must not underestimate its complexity. For any data migration, especially if you are moving off one platform and onto a new one, such as Jira, there are a number of factors you need to consider and prepare for. The following list summarizes some of the common tasks for most data migrations:

- Evaluate the size and impact. This includes the number of records you will be importing and also the number of users that will be impacted by this.
- Perform a full gap analysis between the old system and Jira; for example, consider how the fields will map from one to the other.
- Set up test environments for you to run test imports on to make sure that you have your mappings done correctly.
- Involve your end users as early as possible and have them review your test results.
- Prepare and communicate any outages and support procedures post-migration.

The HR project

Now that we have seen all the key aspects that make up a project, let's revisit what you have learned so far and put it into practice. In this exercise, we will set up a project for our **Human Resource (HR)** team to better track and manage employees joining and leaving the company, as well as tasks related to the recruitment process.

Creating a new project

We will first start by creating a new project for the HR team. To create the project, go through the following steps:

1. Bring up the **Create project** dialog by selecting the **Create project** option from the **Projects** drop-down menu.
2. Select the **Task management** project template. We can use other templates in the **Business** project type; the **Task management** template is the simplest option and will make future customization easier.
3. Name our new project Human Resource and accept the other default values for **Key** and **Project Lead.**
4. Click on the **Submit** button to create the new project.

You should now be taken to the **Project Browser** interface for the new project.

Creating new components

Now that our new project in in place, we need to go ahead and create a few components. These components will serve as groupings for our tasks. We need to perform the following steps to create our new components:

1. Click on the **Project settings** option in the bottom-left corner
2. From the **Project Administration** interface, select the **Components** tab
3. Enter Employee Onboarding for the new component's name
4. Provide a short description for the new component
5. Select a user to be the lead of the component
6. Click on **Add** to create the new component
7. Add a few more components

With projects created as the **Business** project type, components are not displayed by default, so we will have to manually add the **Components** field to the appropriate screens. We will cover fields and screens in Chapter 5, *Field Management*, and Chapter 6, *Screen Management*, respectively. For now, you need to perform the following steps to get our components to display when we create, edit, and view tasks:

1. From the **Project Administration** interface, select the **Screens** tab. There should be three screens, as shown in the following screenshot:

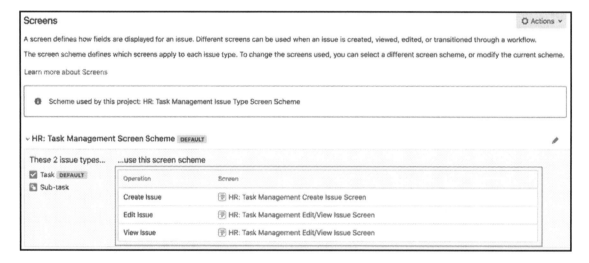

2. Click on **HR: Task Management Create Issue Screen**. This will open the **Configure Screen** page, with a list of fields that are currently on the selected screen.
3. Enter and select **Component/s** in the select field in the bottom of the page; this will add the **Components** field to the screen.
4. Repeat steps 2 and 3 for **HR: Task Management Edit/View Screen**.

Putting it together

Now that you have fully prepared your project, let's see how everything comes together by creating an issue, as follows:

1. Click on the **Create** button from the top navigation bar. This will bring up the **Create Issue** dialog box.

2. Select `Human Resource` for **Project** and **Task** for **Issue Type**.
3. Fill in the fields with some dummy data. Note that the **Component/s** field should display the components we just created.
4. Click on the **Create** button to create the issue.

If everything is done correctly, you should see a dialog box similar to the following screenshot, where you can choose your new project to create the issue in, and also the new components that are available for selection:

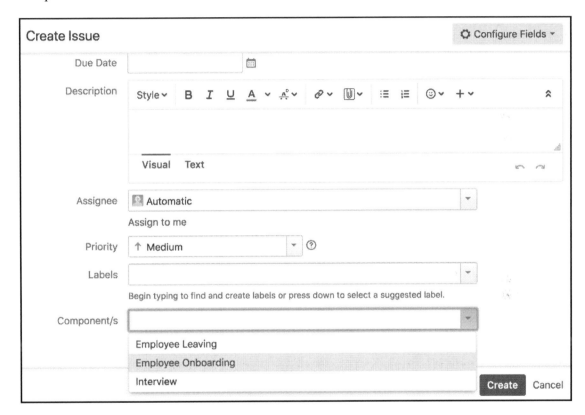

You can test out the default assignee feature by leaving the **Assignee** field as **Automatic** and selecting a component; Jira will automatically assign the issue to the default assignee defined for the component. If everything goes well, the issue will be created in the new project.

Summary

In this chapter, we looked at one of the most important concepts in Jira, projects, and how to create and manage them. Permissions were introduced for the first time, and we looked at three permissions that are related to creating and deleting, administering, and browsing projects.

We were also introduced to the two interfaces Jira provides for project administrators and everyday users—the **Project Administration** interface and **Project Browser** interface, respectively. In the next chapter, we will look at projects created using the Software project type, to enable Scrum and Kanban to run agile projects.

Using Jira for Agile Projects 3

In the previous chapter, we looked at using Jira for normal business projects, a feature that is provided by Jira Core and is also available in Jira Software. In this chapter, we will focus on the two project templates that are exclusive to Jira Software, which are Scrum and Kanban. We will look at a brief overview of each of the agile methodologies, and look at how you can use Jira for both.

By the end of this chapter, you will have learned about the following topics:

- Jira Software project templates
- How to run a project using Jira's Scrum support
- Grooming and managing your backlog
- Estimating work with Scrum
- How to run a project using Jira's Kanban support
- Identifying inefficiencies in your process with Kanban
- Customizing your Scrum and Kanban boards

Scrum and Kanban

Scrum and Kanban are the two agile software development methodologies that have been supported in Jira through an add-on called Jira Agile. Starting with Jira 7, Atlassian has added this support into their Jira Software offering, making agile support a first-class citizen in the product.

If you are already familiar with Scrum and Kanban, feel free to skip this section. However, if you come from a more traditional waterfall model and are new to the agile methodologies, then here is an overview of them both. I would strongly recommend that you pick up an additional resource to learn more about each of the methodologies. A good place to start is the Kanban Scrum minibook, which can be found at https://www.infoq.com/minibooks/kanban-scrum-minibook.

Scrum

Scrum is different to the waterfall model in that it prescribes the notion of iteration. With Scrum, a project is divided into a number of iterations, called **sprints**, each lasting between two to four weeks, with the goal of producing a fully tested and potentially shippable product at the end.

At the beginning of each sprint, the product owner and the team come together in what is called a sprint-planning meeting. In this meeting, the scope of the next sprint is decided. This usually includes top priority items from the backlog, which contains all incomplete work.

During each sprint, the team meets on a daily basis to review progress and flag any potential problems or impediments, and plans how to address them. These meetings are short, and the goal here is to make sure that everyone on the team is on the same page.

At the end of the sprint, the team will come together to review the outcome of the sprint and look at the things they did right and the things that did not go well. The goal is to identify areas of improvement, which will feed into future sprints. This process continues until the project is completed.

Kanban

Kanban, unlike Scrum, which runs in iterations, focuses more on the actual execution of the delivery. It has a heavy emphasis on visualizing the delivery workflow from start to finish, places limits on different stages of the workflow by controlling how many work items are allowed to be in each stage, and measures the lead time.

With Kanban, it is important to be able to visually see the work items going through the workflow, identify areas of inefficiency and bottlenecks, and correct them. It is a continuous process, with work coming in from one end and going out from the other, making sure that things go through as efficiently as possible.

Running a project with Scrum

Scrum is the first agile methodology we will look at. With Jira, a Scrum project consists mainly of two components—the backlog where you and your team will do most of your planning, and the active sprint agile board, which your team will use to track the progress of their current sprint.

Creating a Scrum project

The first step to work with Scrum in Jira is to create a project with the Scrum template. We do this by going through the following steps:

1. Select the **Create project** option from the **Projects** drop-down menu.
2. Choose the **Scrum software development** template and click on **Next**.
3. Accept the settings and click on **Next**.
4. Enter the name and key for the new project and click on **Submit,** as shown in the following screenshot:

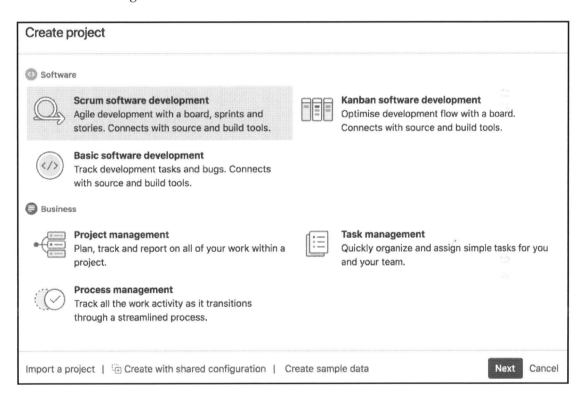

Create project

Software

Scrum software development
Agile development with a board, sprints and stories. Connects with source and build tools.

Kanban software development
Optimise development flow with a board. Connects with source and build tools.

Basic software development
Track development tasks and bugs. Connects with source and build tools.

Business

Project management
Plan, track and report on all of your work within a project.

Task management
Quickly organize and assign simple tasks for you and your team.

Process management
Track all the work activity as it transitions through a streamlined process.

Import a project | Create with shared configuration | Create sample data   Next Cancel

Once the new **Scrum** project has been created, you will be taken to the Scrum interface, as shown in the following screenshot:

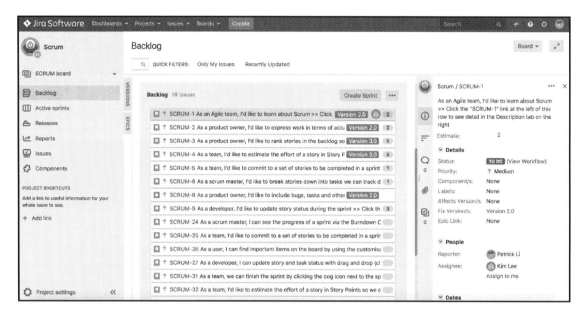

The **Scrum** interface has the following main sections:

- **Backlog**: This is where all unplanned issues are stored. You can think of it as a to-do list. The product owner and the development team will work together to define the priorities of issues in the backlog, which will then be scheduled into sprints for delivery.
- **Active sprints**: This view shows the sprints that are currently underway and the issues that are part of the sprints. This is what the development team will use on a day-to-day basis to track their progress.
- **Reports**: This view contains a number of user reports that you can generate based on your team's performance. These reports help you and your team visualize how the project is progressing and provide valuable feedback that you can use in future sprint-planning sessions for improvements.

Working with the backlog

The backlog is the to-do list of your project, and is where you keep all of your incomplete features (usually represented as stories). When you first start, you may have an empty backlog, so the first step is for the product owner and the team to work together to populate it with stories and tasks that need to be completed. During this step, it works more like a brainstorming session, where the team works together to translate requirements from customers and other stakeholders into actionable stories and tasks.

Prioritizing and estimating work

Once you have the backlog populated, the next step is to estimate and prioritize the issues so that you can plan and build a schedule of how to complete them. In Jira, prioritizing issues in the backlog means moving them up and down in the backlog by dragging and dropping. To increase the priority of an issue, you can simply drag it higher in the backlog list. While it is usually the product owner's responsibility to prioritize which features to deliver first, the team should also be involved in this process to make sure that everyone is in sync regarding the direction of the project.

Estimating work is a critical part of Scrum, and the success of your sprints heavily depends on how well you and your team can estimate. One thing people often get confused about is that they tend to think of estimation in terms of time; for example, story A will take 5 hours to complete and story B will take 10 hours. While this may seem right at first, what would often end up happening is that people will either work extra hard to make the estimate seem accurate or give big estimates because they are uncomfortable with the task at hand. This can lead to big problems as the project goes on, since nobody wants to be known as either the person that cannot give a reliable estimate, or the person that is inefficient because he/she constantly goes over estimates.

One way to avoid this pitfall is to use something arbitrary for estimation called story points, which is the default estimation method in Jira. The idea behind this is to measure and estimate issues based on their complexity, rather than the time required to complete them. So if you start a sprint with a total of 10 story points worth of issues, and at the end of the sprint you are unable to complete all of them, this might indicate that you have been too aggressive and may need to reduce your expectations. Since the estimation is not made based on the time taken, it simply indicates that perhaps the issues are too complex, and you will need to break them down further into more digestible pieces. This helps to prevent people from feeling like they are constantly running against the clock, and also helps you better define and break down tasks into smaller, more manageable chunks.

Sometimes, however, you might find it difficult to estimate the complexity of your stories. This is usually a sign that you either do not have enough information about the story or that the story's scope is too big and needs to be broken down. It is important for the team to realize this and not shy away from going back to ask more questions so that they can fully understand the purpose of the story before providing an estimate.

Now that we have determined a way to estimate our issues, we can move on. Entering the estimate is as simple as doing the following:

1. Select the issue to estimate from the backlog.
2. Enter the number of story points into the **Estimate** field, as shown in the following screenshot:

You should not change the estimate once the issue has been added to an active sprint. Changing the estimate mid-sprint can lead to bad estimation during the spring planning phase and future improvements.

Creating a new sprint

With the backlog populated and issues estimated, the next step is to create a sprint for the team to start working on. To create a new sprint, go through the following steps:

1. Go to the backlog of your project.
2. Click on the **Create Sprint** button. This will create a new empty sprint.
3. Add issues to the sprint by dragging and dropping issues into the sprint box, as shown in the following screenshot:

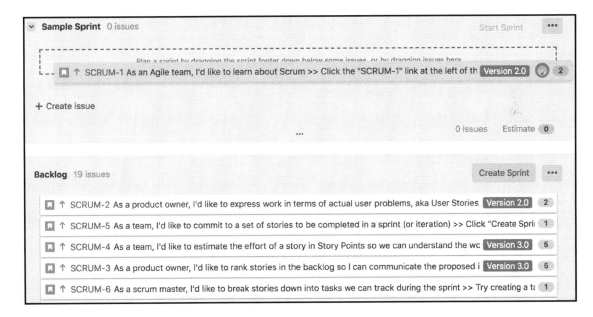

Once the team has decided on the scope, it's time to start the sprint by going through the following steps:

1. Click on the **Start Sprint** button.
2. Select the duration of the sprint. Generally speaking, you want to keep your sprint short. Between one to two weeks is usually a good length.

3. Click on the **Start** button to start your sprint, as shown in the following screenshot:

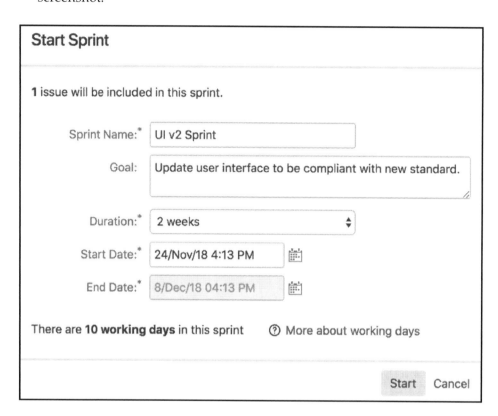

You can select **Custom** for the **Duration** if you want to specify the end date of the sprint yourself instead of using the auto-calculated date.

Once the sprint is started, you can go to the active sprints view and the team can start working on the delivery.

If the **Start Sprint** button is grayed out, it means that you already have an active sprint running and do not have the **parallel sprints** option enabled, or you do not have the Manage Sprints permission. Permissions will be covered in Chapter 9, *Securing Jira*.

Normally, you will only have one team working on the project at any given time, but if you have a big team, and people can work on different parts of the project at the same time, you need to enable the **parallel sprints** option:

1. Log in to Jira as an administrator
2. Browse the Jira administration console
3. Select the **Applications** tab and then **Jira Software configuration**
4. Check the **Parallel Sprints** option to enable it

With the **parallel sprints** option enabled, you can start multiple sprints at the same time. When running multiple sprints, it is best to keep them separate from each other so that they don't get in each other's way. A good example is having two sprints focusing on different areas of the project, such as delivery and documentation.

When you have parallel sprints, since the active sprint view (see the next section) will only show one sprint at a time, you will need to toggle between the sprints, as shown in the following screenshot:

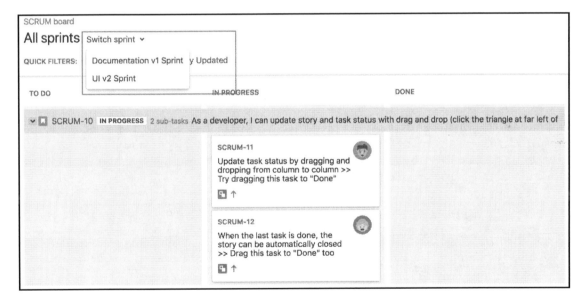

Running through a sprint

Once the team has prioritized the issues and started a sprint during the sprint-planning meeting, the agile board will switch over to the active sprint view. For normal teams, you will have one active sprint at any given time, and your **SCRUM board** will look as follows:

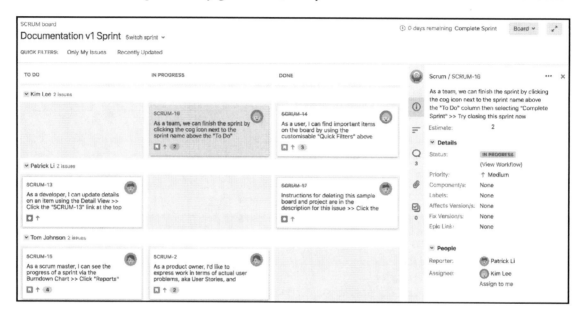

On the **SCRUM board**, each issue is represented as a card, and the board itself is made up of vertical columns that represent the various states an issue can be in, and they are mapped to the workflow that's used by the project. So, in our example, we have the default workflow with three states, **To Do**, **In Progress**, and **Done**. As we will see later, the project administrator will be able to customize this. Team members move the issue card across the board into the appropriate columns as they work and complete their tasks.

The board can also be divided into a number of horizontal rows called **swimlanes**. These help you group similar issues and make your board easier to understand. In our example, we are grouping issues into swimlanes based on the issue assignee. Just like columns, the project administrator can customize how swimlanes should be defined.

When a sprint is underway, it is important to avoid introducing scope creep by adding more issues to the sprint, and it falls on the Scrum master and the product owner to make sure that the team is not distracted or blocked by any impediments. However, from time to time, emergency situations that demand certain features or fixes that need to be included do arise, and in these cases, you can add new issues into the active sprint from the backlog view.

Do keep in mind, though, that this should not become a common habit, as it is a distraction, and it is usually a sign of bad sprint planning or poor communication with stakeholders. For this reason, Jira will prompt you whenever you try to add more issues to an active sprint, as shown in the following screenshot:

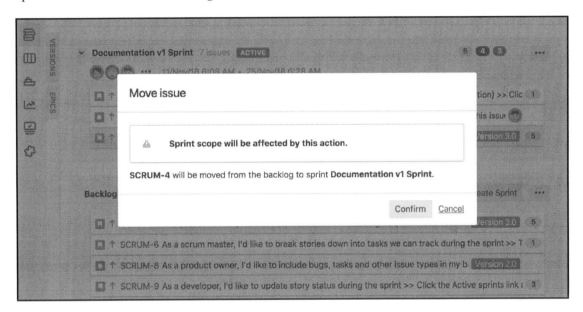

Jira will also show you the scope creep changes in the sprint report if you do add more issues to the sprint.

At the end of the sprint, you need to complete the sprint by doing the following:

1. Go to the **SCRUM board** and click on **Active sprints**.
2. Click on the **Complete Sprint** link, as shown in the following screenshot:
3. Click on the **Complete** button to finish the sprint.

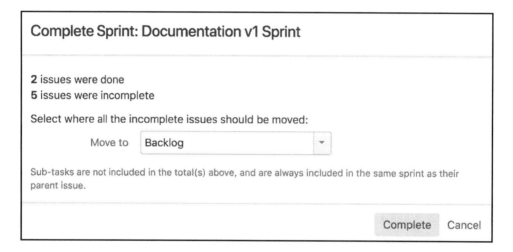

Once you have completed a sprint in Jira, any unfinished issues will be placed back into the backlog. Sometimes, you may have other sprints that are planned but not active; in this case, issues that are not completed from the current active sprint can be automatically added to the next available sprint.

It can be tempting to extend the sprint for just a few more days because you have just one more issue to complete. While this is not a hard rule, you should generally avoid this and just let the incomplete issue go back to the backlog and reprioritize it during the next sprint meeting. This is because Scrum is an iterative process, and the goal is not to make everyone work as hard as possible, but to be able to retrospectively look at what the team did right and/or wrong in the previous sprint and address that in the next sprint. Perhaps the reason for this is because of an inaccurate estimation or incorrect assumptions that were made during requirement gathering. The point here is that the team should view this as an opportunity to improve rather than a failure to be rushed through. Simply extending the current sprint to accommodate incomplete items can turn into a slippery slope where the practice of extending sprints becomes the norm and the root problem is masked.

Running a project with Kanban

Now that we have seen how to run projects with Scrum, it is time to take a look at the other agile methodology Jira Software supports—Kanban. Compared to Scrum, Kanban is a much simpler methodology. Unlike Scrum, which has a backlog and requires the team to prioritize and plan their delivery in sprints, Kanban focuses purely on the execution and measurement of throughput.

In Jira, a typical Kanban board will have the following differences compared to a **SCRUM board**:

- There is no backlog view by default. Since Kanban does not have a sprint-planning phase, your board acts as the backlog. We will see how to enable backlog for Kanban in later sections.
- There are no active sprints. The idea behind Kanban is that you have a continuous flow of work.
- Scrum and Kanban have different type of reports that are specifically designed for each of the methodologies.
- Columns can have minimum and maximum constraints.
- Columns will be highlighted if the constraints are violated. As we can see in the following screenshot, both the **Selected for Development** and **In Progress** columns are highlighted because of constraint violation:

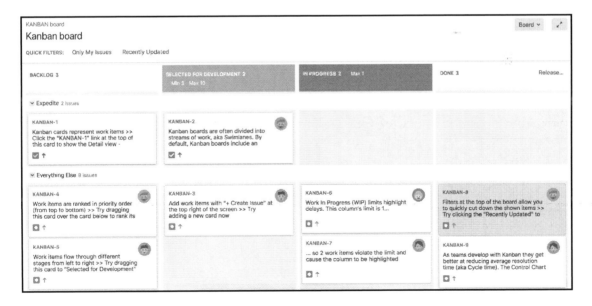

Creating a Kanban project

The first step to working with Kanban in Jira is to create a project with the Kanban template by going through the following steps:

1. Select the **Create project** option from the **Projects** drop-down menu.
2. Choose the **Kanban software development** template and click on **Next,** as shown in the following screenshot:

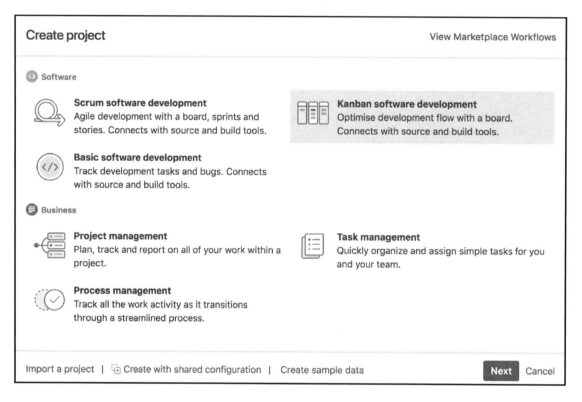

3. Accept the settings and click on **Next**.
4. Enter the name and key for the new project and click on **Submit**.

After you have created a Kanban project, you will be taken to the Kanban board view, which looks very similar to the active sprint view of a **SCRUM board**. Remember, with Kanban, it is as if you are running a sprint that does not end, or that ends when the entire project is completed. Therefore, the agile board itself focuses on helping you and your team to execute the delivery.

Using the Kanban board

As we mentioned earlier, with Kanban, there is by default no sprint planning, so instead of having a view for backlog grooming, everything happens on the **Kanban board** directly. Working with the **Kanban board** is very simple; newly created issues are added to the first column of the board directly, named **Backlog** (by default), as it acts as your backlog of issues with Kanban. Members of the team will then grab issues from the Backlog column, assign the issue to them, and move them through the workflow. During various stages, issues may need to be reassigned to other users—for example, when an issue leaves the development stage and enters testing, it may be reassigned to a test engineer. As more and more issues are completed, you can configure the board to automatically take completed issues off the board after a period of time or perform a release, which will take all issues in the **Done** column from the board (still in the system). The first option is good for teams using Kanban for general task management, and the option to use releases fits better with software development where versions can be released.

Let's look at an example of the **Kanban board**, as shown in the following screenshot, in which we can clearly see that we have problems in both the **In Development** and **In Testing** phases of our process. **In Development** is highlighted in red, meaning that we have enough work there, which is a sign of a bottleneck. **In Testing** is highlighted in yellow, which means that we do not have enough work, which is a sign of efficiency:

With this, the board is able to visually tell us where we are having problems, which allows us to focus on these problem areas. The bottleneck in the **In Development** phase could mean that we do not have enough developers, which causes the efficiency in the **In Testing** phase, where our testers are simply sitting around waiting for work to come through.

This raises a common question—what should be the correct constraints for my columns? The quick answer is, *try and experiment as you go.*

The long answer is, there is no single correct, silver bullet answer. What you need to understand is that there are many factors that can influence the throughput of your team, such as the size of your team, a team member leaving or joining, and the tasks at hand. In our example, the easy solution would be to lower the limit for both columns, and then we are done. But often, it is just as important for you to find the root cause of the problem rather than trying to simply fix the board itself. Perhaps what you should try to do is get more developers on your team so that you can keep up the pace that is required for delivery. The take away here is that the **Kanban board** can help you pinpoint the areas of a problem, and it is up to you and your team to figure out the cause and find the appropriate solution.

Enabling the backlog for the Kanban board

For those of you who come from Scrum, not having a proper backlog may feel uncomfortable, or perhaps as your project grows, having all the new issues being displayed on the **Kanban board** in the **Backlog** column becomes too unwieldy. The good news is, Jira supports a hybrid methodology of Kanban and Scrum, called **Kanplan**, which lets you have a backlog for Kanban as well.

To add a Scrum style backlog to your Kanban project, simply map the appropriate status into the Kanban backlog column by going through the following steps:

1. Browse to your project's agile board.
2. Click on the **Board** menu and select the **Configure** option.
3. Select the **Columns** option from the left navigation panel. By default, there is a column called **Backlog**.
4. Delete the **Backlog** column.
5. Drag and drop the now unmapped **Backlog** status from the **Unmapped Statuses** column on the right to the **Kanban backlog** on the left, as shown in the following screenshot:

With a status mapped in the **Kanban backlog** column, you will now have a backlog that looks and works just like a Scrum project, and any newly created issues will be added to the backlog since they will be in the **Backlog** workflow status, as shown in the following screenshot:

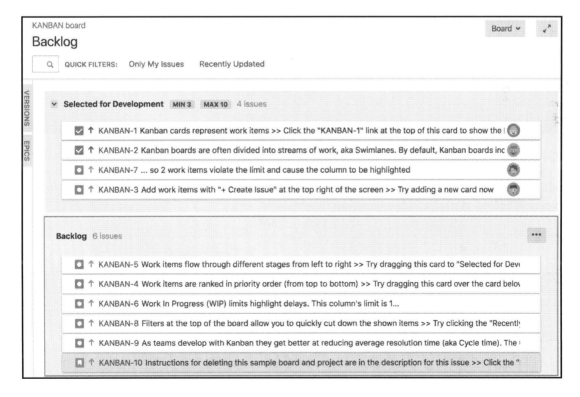

You can map more than one status to the backlog, and as we will see in `Chapter` `7`, *Workflow and Business Process*, when we cover workflows in depth, you might have a more complex process, and issues in different statuses can appear in the backlog. Another benefit of having a backlog for Kanban is that if you have hundreds of issues, it is a lot faster and more efficient for Jira to display them this way than on the **Kanban board**, and Jira 8 has made significant improvements to the backlog's performance.

Configuring agile boards

Now that we have seen how to use Jira Software to run both Scrum and Kanban projects, let's take a look at how to customize our agile board. Since Jira Software is built on top of Jira Core, many of its customization options leverage the core features of Jira. If you are not familiar with some of these features, such as workflows, don't worry—we will cover these at a high level in the context of the agile board and dive into the details of each in later chapters.

Configuration columns

For both Scrum and Kanban, the board's columns are mapped to the workflow that's used by the project, and the default workflow that's created is very simple. For example, the default Scrum workflow contains three statuses—**To Do**, **In Progress**, and **Done**. However, this is often not enough, as projects will have additional steps in their development cycle, such as testing and review. To add new columns to your board, follow these steps:

1. Browse to your project's agile board.
2. Click on the **Board** menu and select the **Configure** option.
3. Select the **Columns** option from the left navigation panel.
4. Click on the **Add Column** button.
5. Enter the name for the new column and select its category. Generally speaking, your new column will fall into the **In Progress** category, unless you are replacing the **To Do** or **Done** column.
6. Drag and drop the new column to place it in the correct location within your development workflow.

For projects that are using the workflows that were created along with your new project (also known as a **Simplified Workflow**), this is all you need to do to customize your columns, as shown in the following screenshot. If you have an existing workflow and want to adapt your columns to that, then you will learn how to do this when we cover workflows in Chapter 7, *Workflow and Business Process:*

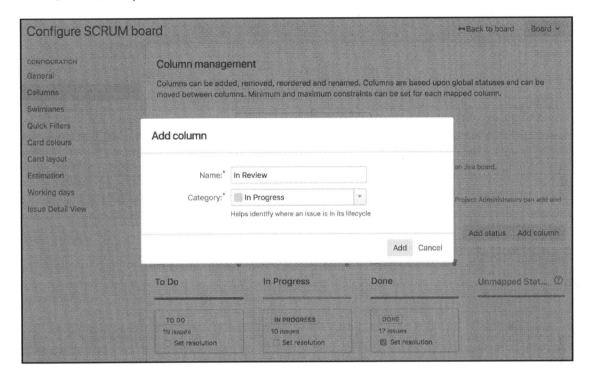

Setting up column constraints

In the previous *Kanban* section, we mentioned that one of the key aspects of Kanban is to control the amount of work that is sent through.

> While a work constraint is a concept that's used in Kanban, sometimes people also adopt it with Scrum. This allows you to use Scrum for planning and Kanban for execution in a hybrid methodology called **Scrumban**.

To set up column constraints for your agile board, go through these steps:

1. Browse to your project's agile board.
2. Click on the **Board** menu and select the **Configure** option.
3. Select the **Columns** option from the left navigation panel.
4. Select how you want the constraint to be calculated in the **Column Constraint** option. By default, Kanban board will use the **Issue Count** option, while the **Scrum board** will not have any constraint.
5. Enter the minimum and maximum value for each of the columns you want to apply a constraint to.

You do not have to set both the minimum and maximum for a constraint. Consider the example that's shown in the following screenshot, where we have set the constraint for **Selected for Development** to have at least three issues and no more than ten issues. For the **In Progress** column, we have only limited it to be no more than five issues, but there is no minimum value, meaning that it can have no issues at all. We also placed a maximum limit of 15 issues for the **Done** column so that we are alerted when the team has reached the threshold of completed issues and a release needs to be made:

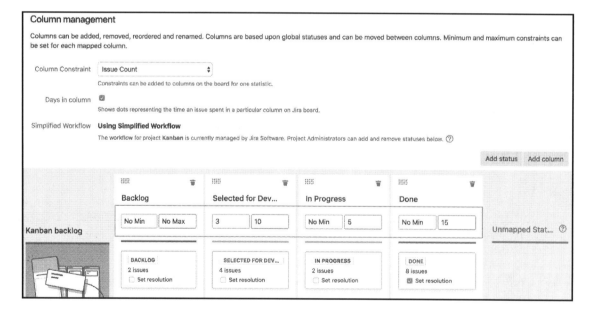

After you have set the column constraints for your board, every time the rules are violated, Jira will immediately alert you on your agile board. For example, in the following screenshot, we have two issues in the **Selected for Development** column, which has a minimum of three issues, so the column is highlighted in yellow. In the **In Progress** column, we have six issues, and since it has a maximum limit of 5 issues, the column is highlighted in red.

Note that while Jira will highlight the columns when a constraint is violated, it does not actually stop you from breaking the constraints. It is simply a way to alert the team that something has gone wrong in the process and that it needs to be reviewed and corrected:

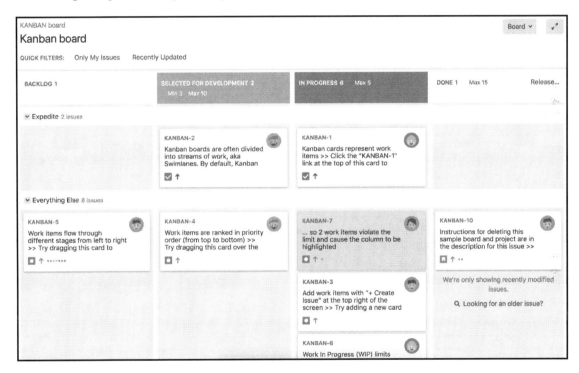

Configuring swimlanes

As we saw in earlier sections, Jira's agile board lets you group similar issues together in horizontal rows called swimlanes. Unlike columns, which are mapped to workflow statuses, you can define swimlanes based on any criteria, including custom fields you have added yourself. To set up swimlanes for your board, you need to go through the following steps:

1. Browse to your project's agile board.
2. Click on the **Board** menu and select the **Configure** option.
3. Select the **Swimlanes** option from the left navigation panel.
4. Select how you want to define your swimlanes in the **Base Swimlanes on** field.
5. If you choose the **Queries** option, you will need to define the query for each swimlane you want to add to the board.

There are six options to choose from when choosing how swimlanes will be defined, as described in the following list:

- **Queries**: Swimlanes will be based on the **Jira Query Language** (**JQL**) queries you define. For each swimlane, you need to define the JQL query that will return the issues you want for the swimlane. Issues that match more than one query will only be included in the first swimlane. JQL will be covered in `Chapter 10`, *Searching, Reporting, and Analysis*.
- **Stories**: Swimlanes will be based on user stories. Subtasks that belong to the same story will be displayed in the same swimlane.
- **Assignees**: Swimlane will be based on each issue's assignee. Issues with the same assignee will be grouped in the same swimlane. The sample Scrum board we have shown in the *Scrum* section uses this option.
- **Epics**: Swimlanes will be based on epics that each issue belongs to. Issues in the same epic will be grouped into the same swimlane.
- **Projects**: Swimlanes will be based on the project each issue belongs to. As we will see later in this chapter, an agile board can include issues from multiple projects.
- **No Swimlanes**: The agile board will not be using swimlanes, so all issues will be grouped together in one single row.

As we can see in the following screenshot, we are using the **Queries** option and we have defined two swimlanes (and the default **Everything Else** lane). For the JQL query, we are searching based on a custom field that we have created called **Source** to determine whether the feature request comes from a customer or as a result of an internal review. Custom fields will be covered in `Chapter 5`, *Field Management*.

Using queries is the most flexible option when it comes to configuring your swimlanes, since Jira's query language is very powerful and allows you to define any arbitrary rule for your **Swimlanes**:

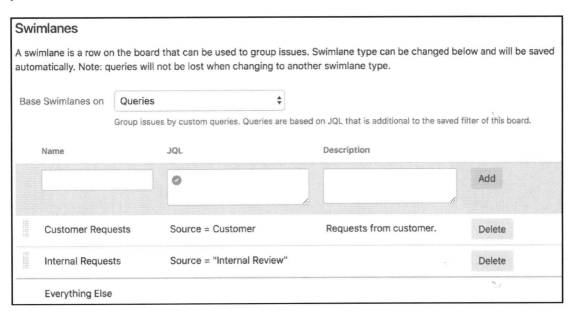

Defining quick filters

By default, your agile board will display all issues. For Scrum, it will be all issues in the sprint, and for Kanban, it will be all issues that have not been released. This can be quite distracting when you have many issues and only want to focus on specific ones. While swimlanes can help with that, having too many issues on the board can still be very *noisy*.

One useful feature Jira has is that you can create a number of predefined filters for your board. With these, you can quickly filter out the issues you do not care about and only have the issues that matter to you shown on the board. Note that this means that the other issues aren't removed from the board—they are simply hidden from view.

Jira already comes with two built-in quick filters, called **Only My Issues** and **Recently Updated**. You can create your own by following these steps:

1. Browse to your project's agile board
2. Click on the **Board** menu and select the **Configure** option
3. Select the **Quick Filters** option from the left navigation panel
4. Enter a name for the new filter and the JQL query that will return the filtered issues

In the following screenshot, we are creating a new filter called **Customer Requests** and using the JQL to search for issues with the **Source** field set to **Customer**:

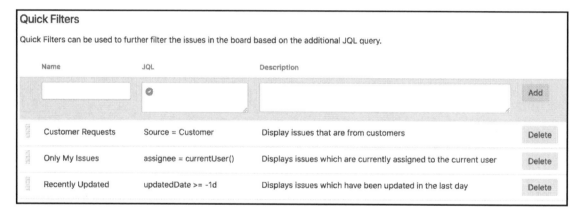

After you have added your new filter, it will be displayed next to the existing ones, ordered alphabetically. Clicking on the filter will immediately filter out the issues that do not fit the criteria. You can also chain filter the issues by selecting multiple filters, as shown in the following screenshot. Note that this is combining the selected filters using an AND, not OR. We have enabled both the **Customer Requests** and **Recently Updated** filters so that we can get a view on recently updated customer requests:

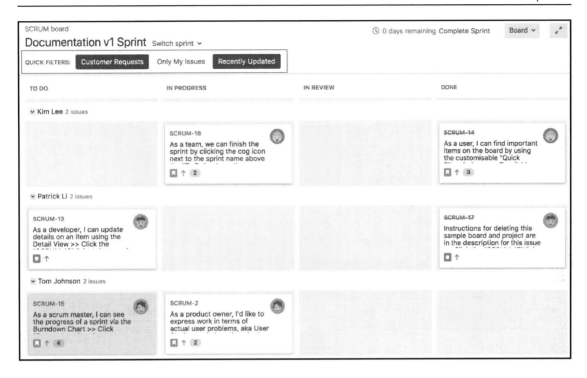

Grooming your backlog

If you are using Scrum or Kanplan, then a big part of your routine will be to groom your backlog. This means making sure that high priority items are floated to the top and not getting buried. This is a constant exercise and is especially important as you and your team approach the start of a new sprint, in the case of Scrum. With Kanplan, it is just as important to prioritize the tasks so that your team can maintain their throughput and not violate any constraints because of poor planning.

Jira's backlog comes with several handy features to help you avoid turning backlog grooming into a tedious exercise. To prioritize issues in your backlog, you simply move the high priority issues up and move the low priority issues down. While this seems very simple, as your backlog grows, it can become tricky to drag an issue all the way from the bottom of a backlog with over hundreds of issues to the top, and let's not forget that newly added issues go to the bottom by default. What you can do in this case is right-click the issue you want to move and select the **Top of Backlog** option from the **Send to** menu. This will move the issue to the top of the backlog. You can do this with multiple issues as well by simply hold down the *Shift* or *Ctrl* key on your keyboard to select all the issues you want to move and either use the **Send to** menu or drag and drop, as shown in the following screenshot:

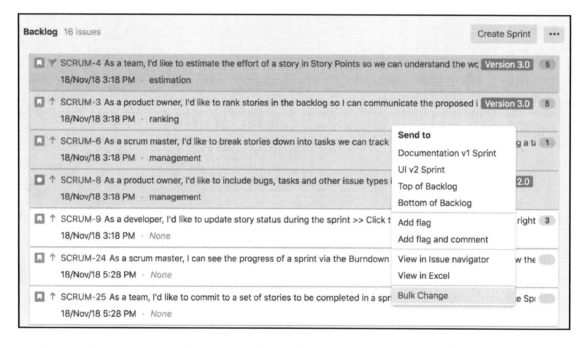

Another useful tool is the ability to flag issues that are important or require special attention, which is done by right-clicking an issue and selecting the **Add flag** option. Once an issue has been flagged, a flag icon will appear next to the issue, and will always be highlighted. Note that this is mostly for visual purposes only; Jira will not force you to immediately work on a flagged issue in any way.

Lastly, you can display additional information about the issues in the backlog. These can be data from the out-of-box fields, or any custom fields you create. A common use case for this is to display the labels that have been added to issues, as a label is an easy way for users to add a tag to an issue for various purposes. To add more field values to the issue card in the backlog, follow these steps:

1. Browse to your project's agile board.
2. Click on the **Board** menu and select the **Configure** option.
3. Select the **Card layout** option from the left navigation panel.
4. Select the field you want to add from under the **Backlog** section. You can add up to three fields.

As we can see in the preceding screenshot, we have added both the created date and labels to the issue card in the backlog.

Creating a new agile board for your project

When you create a new project using the **Scrum** and **Kanban** project template, as described earlier in this chapter, Jira will automatically create an agile board for your project. Along with this default board, you can create additional boards for your project.

For example, if you created a **Scrum** project, and you have two teams working on the project, you can create a new **Scrum board** for the second team so that each team can work with their own agile boards and not get in each other's way. Another example is if your second team needs to run their part of the project using **Kanban**, then you can easily add a new **Kanban board** to the **Scrum** project so that each team can use the agile methodology they want for the same project. To add a new agile board to your project, follow these steps:

1. Browse to your project's agile board.
2. Click the current board's name from the top left-hand corner and select the **Create board** option.

3. Select the agile board type you want to create and follow the onscreen wizard to create the new board, as shown in the following screenshot:

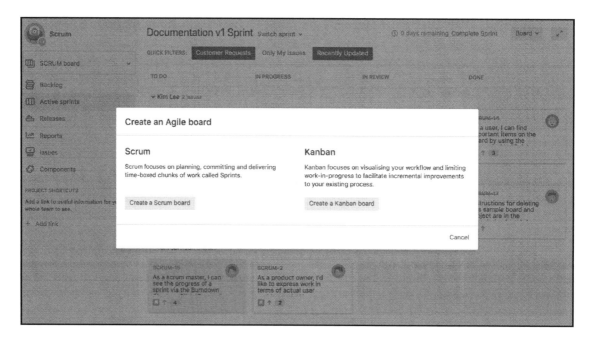

Once the new agile board has been created, it will be added to the agile boards menu from the top left-hand corner, and you can switch between the boards by selecting the one you want.

Including multiple projects on a board

By default, when you create a new project, the agile board that's created will only include issues from the current project. This is usually fine if your project is self-contained; however, there might be cases where you have multiple projects that are related or dependent on each other, and for you to get an overall picture, you need to have issues from all of those projects shown on a single agile board.

The good news is that Jira lets you do just this. One thing to understand here is that Jira uses what is called a **filter** to define what issues will be included on the board. Filters are like saved search queries, and when a project is created, Jira automatically creates a filter that includes all of the issues from the current project. This is why the default agile board that's created with the project will always display the project's issues. Filters will be discussed in Chapter 10, *Searching, Reporting, and Analysis*.

So for you to include issues from other projects on the agile board, all you need to do is update the board using the following steps:

1. Browse to your project's agile board.
2. Click on the **Board** menu and select the **Configure** option.
3. Select the **General** option from the left navigation panel.
4. Click the **Edit Filter Query** link for the **Saved Filter** option if you want to update the filter that is currently being used by the board, as shown in the following screenshot. Usually only the filter owner can change a filter's query, but since filters created with the board has the same owner as the board by default, you are able to edit the query. Alternatively, if you already have a filter that has all the issues you want to include, you can hover over and click on the current filter and then select the new filter to use:

 Since filters need to be shared with users in order for them to see the issues they return, make sure your filter is shared with the same group of users that the board is set to. Generally, you can just share the filter with the project.

General and filter

The Board filter determines which issues appear on the board. It can be based on one or more projects, or custom JQL depending on your needs.

General

Board name	**SCRUM board**
Administrators	**Patrick Li (admin)**

Filter

Saved Filter	**Filter for SCRUM board** ✎ — click to select a different filter Edit Filter Query
Shares	**No shares** Edit Filter Shares
Filter Query	**project = SCRUM ORDER BY Rank ASC**
Ranking	**Using Rank**
Projects in board	**Scrum** View permission

Summary

In this chapter, we introduced the software project templates that come with Jira Software and the two main agile methodologies it supports, namely Scrum and Kanban. We talked about how you can run projects in each of the methodologies using Jira and the features it provides.

We also looked at some of the customization options that are available for you as the project owner so that you can configure the agile board to fit your needs. We looked at how to customize the board's columns to better adapt to your team's workflow and also looked at using swimlanes to group similar issues together. We also looked at how to create quick filters to easily filter out irrelevant issues from view so that we can focus on the issues that matter.

In the next chapter, we will look at issues, the key data you work with in your projects, and what you can do with it.

Section 2: Jira 8 in Action

In this section, you will learn how to use Jira. You will get acquainted with techniques for handling different issues, exploring fields, creating new screens, managing workflows, and setting up incoming and outgoing email servers.

The following chapters will be covered in this section:

- Chapter 4, *Issue Management*
- Chapter 5, *Field Management*
- Chapter 6, *Screen Management*
- Chapter 7, *Workflow and Business Process*
- Chapter 8, *Emails and Notifications*

Issue Management

4

In the previous chapter, we saw that Jira is a very flexible and versatile tool that can be used in different organizations for different purposes. A software development organization will use Jira to manage its software development life cycle and track bug, while a customer services organization may choose to use Jira to track and log customer complaints and suggestions. For these reasons, issues in Jira can represent anything that is applicable to real-world scenarios. Generally speaking, an issue in Jira often represents a unit of work that can be acted upon by one or more people.

In this chapter, we will explore the basic and advanced features that are offered by Jira so that you can manage issues. By the end of this chapter, you will have learned about the following topics:

- Issues and how they are used in Jira
- Creating, editing, and deleting issues
- Moving issues between projects
- Expressing your interest in issues through voting and watching
- Advanced issue operations, including uploading attachments and linking issues

Understanding issues

Depending on how you are using Jira, an issue can represent different things, and can even look very different in the user interface. For example, in Jira Core, an issue will represent a task, and will look like this, as shown in the following screenshot:

In Jira Software, if you are using the agile board, an issue can represent a story or epic, and will resemble a card:

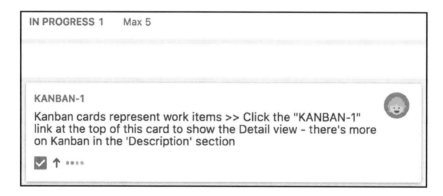

Despite all the differences in what an issue can represent and how it might look, there are a number of key aspects that are common for all issues in Jira, listed as follows:

- An issue must belong to a project.
- It must have a type, otherwise known as an issue type, which indicates what the issue represents.
- It must have a summary. The summary acts like a one-line description of what the issue is about.

- It must have a status. A status indicates where along the workflow the issue is at a given time. We will discuss workflows in `Chapter 7`, *Workflow and Business Process*.

In summary, an issue in Jira represents a unit of work that can be completed by a user, such as a task in Jira Core, a story in Jira Software, or a request in Jira Service Desk. These are all different forms of an issue.

Jira issue summary

As we have already discussed, an issue in Jira can be anything in the real world to represent a unit of work or a task to be completed. In this section, we will look at how Jira presents an issue in the user interface for Jira Core and Jira Software. We will cover Jira Service Desk in `Chapter 11`, *Jira Service Desk*, as it has a different interface.

First, let's take a look at an issue in Jira Core. The following screenshot shows a typical example of an issue and breaks it down into more digestible sections, followed by an explanation of each of the highlighted sections in a table. This view is often called the issue summary or the **View Issue** page:

These sections are described in the following table:

Section	Description
project/issue key	This shows the project the issue belongs to. The issue key is the unique identifier of the current issue. This section acts as a breadcrumb for easy navigation.
issue summary	This is a brief summary of the issue.
issue export options	These are the various view options for the issue. The options include XML, Word, and Printable.
issue operations	These are the operations that users can perform on the issue, such as edit, assign, and comment. These are covered in later sections of this chapter.
workflow options	These are the workflow transitions that are available. Workflows will be covered in Chapter 7, *Workflows and Business Process*.
issue details /fields	This section lists the issue fields, such as issue type and priority. Custom fields are also displayed in this section. Fields will be covered in Chapter 5, *Field Management*.
user fields	This section is specific for user-type fields, such as assignee and reporter. Fields will be covered in Chapter 5, *Field Management*.
date fields	This section is specific for date-type fields, such as create and due date. Fields will be covered in Chapter 5, *Field Management*.
attachments	This lists all the attachments in an issue.
sub-tasks	Issues can be broken down into smaller subtasks. If an issue has subtasks, they will be listed in this section.
comments	This lists all the comments that are visible to the current user.
work log	This lists all the time-tracking information your users have logged against the issue. See the *Time tracking* section for more details.
history	Keeps track of all changes that have occurred for this issue, including the values that are used before and after the change.
activity	This is similar to history, but is formatted in a more user-friendly way. It can also generate an RSS feed for the content.

Jira Software, when running Scrum or Kanban, uses the agile board user interface, which represents issues as cards on a board, as described in Chapter 3, *Using Jira for Agile Projects*, which has a more concise summary of issues. However, when you click on the card, Jira will expand and display the detailed information of the issue, as shown in the preceding table.

Working with issues

As we have already seen, issues are at the center of Jira. In the following sections, we will look at what you, as a user, can do with issues. Note that each of these actions will require you to have specific permissions, which we will cover in Chapter 9, *Securing Jira*.

Creating an issue

When creating a new issue, you will need to fill in a number of fields. Some fields are mandatory, such as the issue's summary and type, while others are optional, such as the issue's description. We will discuss fields in more detail in the next chapter.

There are several ways in which you can create a new issue in Jira. You can choose any of the following options:

- Click on the **Create** button at the top of the screen
- Press *C* on your keyboard

This will bring up the **Create Issue** dialog box, as shown in the following screenshot:

As you can see, there are quite a few fields, and the required fields will have a red asterisk (*) mark next to their names.

Apart from manually creating issues this way, you can also use advanced tools, such as issue importer, Jira's REST APIs, and emails, to create issues.

The administrator configures what fields will be part of the create dialog, but as a user, you can customize and make your own create screen by hiding the optional fields by performing the following steps:

1. Click on the **Configure Fields** option in the top-right corner.
2. Select the **Custom** option.
3. Uncheck all the fields you want to hide and check the fields that you want to display, as shown in the preceding screenshot.

 Make sure you do not hide any required fields, otherwise you will not be able to create new issues. You are only hiding or showing these fields for yourself. Only the Jira administrator can actually hide and show fields globally for all users.

There is a **Create another** option beside the **Create** button. By ticking this option and then clicking on the **Create** button, the **Create Issue** dialog box will stay on the screen and remember the values you have previously entered, such as priorities components, and due dates. This way, you can avoid having to fill in the whole dialog box again and will only have to update some of the fields that are actually different, such as **Summary**. With this feature, you can rapidly create many issues in a much shorter time frame.

Editing an issue

There are two ways in which you can edit an issue in Jira. The first and more traditional way is by clicking on the **Edit** button or by pressing *E* on your keyboard. This will bring up the **Edit Issue** dialog box, along with all the editable fields for the current issue. This allows you to make changes to multiple fields at once.

The second option is called in-line editing. With this feature, you will be able to view the issue and edit the field you want on the spot, without having to wait for the edit dialog to load. Scroll down to find the field. To edit a field inline, all you have to do is hover your mouse over the value for the field you want to update, wait for the edit icon to show up, click on the icon, and start editing, as shown in the following screenshot:

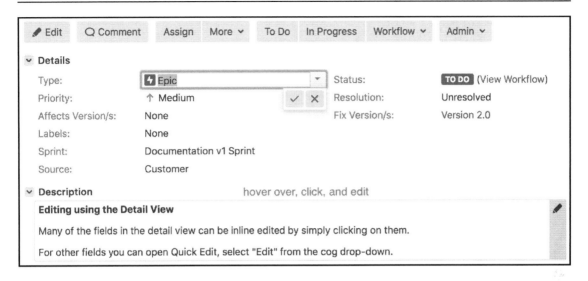

The fields you can edit are controlled by the screen that's used for the edit issue operation. Screens will be discussed in `Chapter 6`, *Screen Management*.

Deleting an issue

You can delete issues from Jira. You might need to delete issues that have been created by mistake or if the issue is redundant, although normally, it is better to close and mark the issue as a duplicate. We will discuss closing an issue in `Chapter 7`, *Workflow and Business Process*.

 Issue deletion is permanent in Jira. Unlike some other applications that may put deleted records in a trash bin that you can retrieve later, Jira completely deletes the issue from the system. The only way to retrieve the deleted issue is by restoring Jira from a previous backup.

Go through the following steps to delete an issue:

1. Browse to the issue you wish to delete.
2. Click on the **Delete** option from the **More** menu. This will bring up the **Delete Issue** dialog box, as shown in the following screenshot:

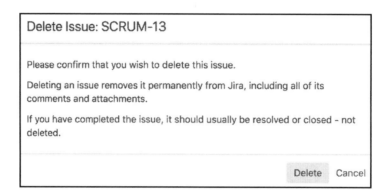

3. Click on the **Delete** button to remove the issue permanently from Jira.

Deleting an issue permanently removes it from Jira, along with all of its data, including attachments and comments.

Moving an issue between projects

Once an issue has been created, the issue is associated with a project. You can, however, move the issue around from one project to another. This may sound like a very simple process, but there are many steps involved and many things that need to be considered.

First, you need to decide on a new issue type for the issue if the current issue type does not exist in the new project. Second, you will need to map a status for the issue if the target project uses a different workflow. Third, you will need to decide on the values for any mandatory fields that exist in the new project but that do not exist in the current project. Sound like a lot? Luckily, Jira comes with a wizard that is designed to help you address all of these things.

Go through the following steps to start moving an issue:

1. Browse to the issue you wish to move.
2. Click on the **Move** option in the **More** menu. This will bring up the **Move Issue** wizard.

There are essentially four steps in the **Move Issue** wizard.

The first step is to select which project you wish to move the issue to. You will also need to select the new issue type, as shown in the following screenshot. If the same issue type exists in the new project, you can usually continue and use the same issue type:

The second step allows you to map the current issue to the new project's workflow, as shown in the following screenshot. If the issue's status exists in the target project, the wizard will skip this step:

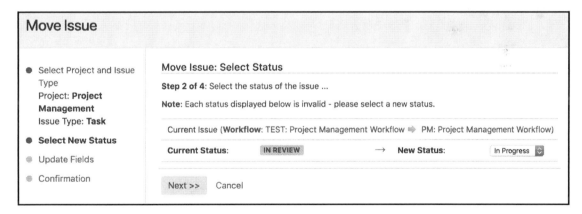

The third step shows all the fields that exist in the new project but not the current project, as well as those that require a value, as shown in the following screenshot. Again, if there are no fields that require values to be set, this step will be skipped:

The fourth and last step shows you the summary of the changes that will be applied by moving the issue from the source project to the target project. This is your last chance to make sure that all the information is correct. If there are any mistakes, you can go back to step one and start over again. If you are happy with the changes, confirm the move by clicking on **Move**, as shown in the following screenshot:

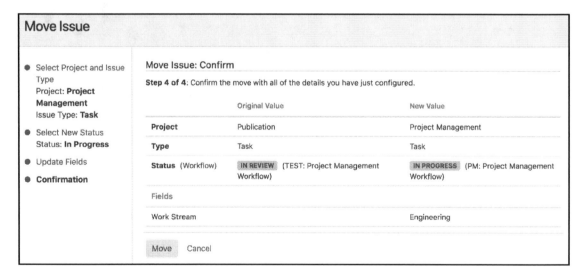

Once the issue has been moved, it will be given a new issue key based on the new project. If you access the issue with its old issue key, Jira will automatically redirect you.

Casting a vote on an issue

The most straightforward way to express your interest in a Jira issue is to vote for it. For organizations or teams that manage their priorities based on popularity, voting is a great mechanism for collecting this information.

An example of this is how Atlassian uses Jira (`https://jira.atlassian.com/browse/JRASERVER-9`) as a way to let its customers choose and vote for the features they want to be implemented or bugs to be fixed by voting on issues based on their needs. This allows the product management and marketing team to have an insight into market needs and how to best evolve their offerings.

One thing to keep in mind when voting is that you can only vote once per issue. You can vote many times for many different issues, but for any given issue, you have only one vote. This helps prevent a single user from continuously voting on the same issue, which may skew the voting result. You can, however, unvote a vote you have already cast on an issue and vote for it again later; if you choose to do this, it will still only count as one vote.

To vote for an issue, simply click on the **Vote for this issue** link next to **Votes**. When you have voted for an issue, the icon will appear as colored. When you have not yet voted for an issue, the icon will appear as grayed out. Note that you cannot vote for issues that you yourself have created.

Receiving notifications on an issue

Jira is able to send automated email notifications about updates regarding issues to users. Normally, notification emails will only be sent out to the issue's reporter, assignee, and people who have registered interest in the issue. This behavior can be changed through notification schemes, which we will discuss in `Chapter 8`, *Emails and Notifications*.

You can register your interest in the issue by choosing to watch the issue. By watching an issue, you will receive email notifications on activities, such as new comments and issue updates. Users watching the issue can also choose to stop watching and thus stop receiving email updates from Jira. You can also add other users as watchers by adding them to the watcher's list.

To watch an issue, simply click on the **Start watching this issue** link. If you are already watching the issue, it will change to **Stop watching this issue**. If you click on the link again, you will stop watching the issue, as shown in the following screenshot:

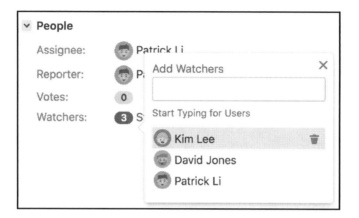

 Jira will automatically add you as a watcher for issues that are created by you, or the issues you have commented on and updated.

Jira also shows you how many people are actively watching the issue by displaying the total watchers next to the watch icon. You can click on the number next to **Watchers** to see the full list of watchers, and add new users as watcher to the issue.

Assigning issues to users

Once an issue has been created, the user that's normally assigned to the issue will start working on it. Afterwards, the user can assign the issue further, for example, to QA staff for further verification.

There are many instances where an issue needs to be reassigned to a different user, for example, when the current assignee is unavailable or if issues are created with no specific assignee. Another example is where issues are assigned to different people at different stages of the workflow. For this reason, Jira allows users to reassign issues once they have been created.

Go through the following steps to assign an issue:

1. Browse to the issue you wish to assign.
2. Click on the **Assign** button in the **Issue** menu bar or press *A* on your keyboard (you can also use the in-line edit feature here). This will bring up the **Assign** dialog.
3. Select the new assignee for the issue and optionally add a comment to provide some information to the new assignee.
4. Click on the **Assign** button, as shown in the following screenshot:

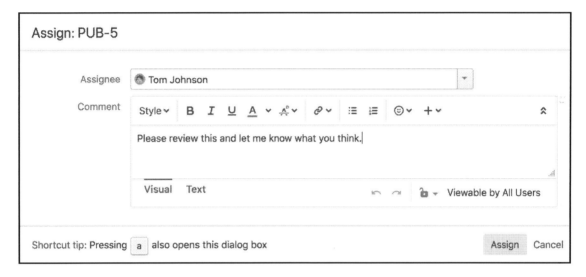

Once this issue has been reassigned, its assignee's value will be updated to the new user. The new assignee will also receive a notification email, alerting them of the assignment. You can also unassign an issue this way by simply selecting the **Unassigned** option. Unassigned issues do not have an assignee and will not show up on anyone's list of active issues.

You can press *I* on your keyboard to quickly assign the issue to yourself.

Sharing issues with other users

If you want to email an issue to other users in Jira, instead of having to manually copy and paste the issue's URL in an email, you can use the built-in share feature in Jira. All you have to do is go to the issue you want to share and click on **Share**, as shown in the following screenshot, or press *S* on your keyboard. Then, select the users you want to share the issue with and click on the **Share** button, as shown in the following screenshot:

 If the user you are sharing the issue with does not have access to the issue, they will not be able to see the issue's details.

Issue linking

Jira allows you to create custom hyperlinks for issues. This allows you to provide more information about the issue. There are two types of links you can create—linking to other issues in Jira or linking to any arbitrary resources on the web, such as a web page.

Linking issues with other issues

Issues are often related to other issues in some way. For example, issue A might be blocking issue B, or issue C might be a duplicate of issue D. You can add descriptions to the issue to convey this information, or delete one of the issues in the case of duplication, but with this approach, it is hard to keep a track of all of these relationships. Luckily, Jira provides an elegant solution for this with the standard issue link feature.

The **standard issue link** lets you link an issue with one or more other issues in the same Jira instance. This means that you can link two issues from different projects together (if you have access to both the projects). Linking issues in this way is very simple; all you need to know is the target issues to link to. You can link issues by going through the following steps.:

1. Browse to the **View Issue** page for the issue you wish to create a link for.
2. Select **Link** from the **More** menu. This will bring up the link issue dialog box.
3. Select the **Jira Issue** option from the left panel.
4. Select the type of issue linking from the **This issue** drop-down menu.
5. Select the issues to link to. You can use the search facility to help you locate the issues you want.
6. Click on the **Link** button, as shown in the following screenshot:

After you have linked your issues, they will be displayed in the **Issue Links** section on the **View Issue** page. Jira will display the target issue's key, description, priority, and status.

Linking issues with remote contents

The standard Jira issue link allows you to link multiple issues to the same Jira instance. Jira also lets you link issues to resources, such as web pages on the internet.

Using remote issue links is quite similar to the standard issue link; the difference is that instead of selecting another issue, the URL address of the target resource is specified. You can set this up by going through the following steps:

1. Open the **Link Issue** dialog box.
2. Select the **Web Link** option from the left panel.
3. Specify the URL address for the target resource. Jira will automatically try and find and then load the appropriate icon for the resource.
4. Provide the name for the link in the **Link Text** field. The name you provide here will be what is shown for the link when viewing the issue.
5. Click on the **Link** button, as shown in the following screenshot:

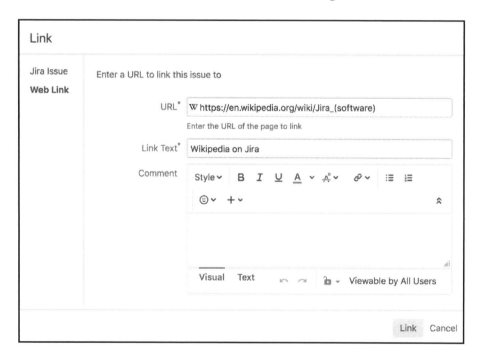

Issue cloning

When you need to create a new issue and you already have a baseline issue, Jira allows you to quickly create it with the data based on your existing issues by cloning the original one. Cloning an issue allows you to quickly create a new one, with most of its fields populated. For example, you might have two software products with the same bug. After creating a bug report in one project, you can simply clone it for the other project.

A cloned issue will have all the fields copied from the original issue; however, it is a separate entity. Further actions performed on either of the two issues will not affect the other.

When an issue is being cloned, a **Clone** link is automatically created between the two issues, establishing a relationship.

Cloning an issue in Jira is simple and straightforward. All you have to do is specify a new summary (or accept the default summary with the text CLONE at the front) for the cloned issue by going through the following steps:

1. Browse to the issue you wish to clone
2. Select **Clone** from the **More** menu
3. Enter a new summary for the newly cloned issue
4. Check the **Clone Sub Tasks** checkbox if you also want to copy over all the subtasks
5. Click on the **Create** button

Once the issue has been successfully cloned, you will be taken to the issue summary page for the newly cloned issue.

Time tracking

Since issues often represent a single unit of work that can be worked on, it is logical for users to log the time they have spent working on it. You can specify the estimated effort that is required to complete an issue, and Jira will be able to help you track its progress.

Jira displays the time tracking information of an issue in the **Time Tracking** panel on the right-hand side, as shown in the following screenshot:

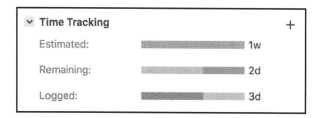

The **Time Tracking** panel includes of the following information:

- **Estimated**: This represents the original estimated effort that's required to complete the issue, for example, the estimated time required to fix a bug by a developer.
- **Remaining**: This represents the remaining time for the issue to be completed. It is calculated automatically by Jira based on the original estimate and total time logged by users. However, the user logging work on the issue, as described in the following section, can also override this value.
- **Logged**: This represents the total time spent on the issue so far.

Specifying original estimates

Original estimates represent the anticipated time required to complete the work that's represented by the issue. It is shown as the blue bar under the **Time Tracking** section.

For you to specify an original estimate value, you need to make sure that the **Time Tracking** field is added to the issue's create and/or edit screen. We will discuss fields and screens in `Chapter 5`, *Field Management*, and `Chapter 6`, *Screen Management*, respectively.

To specify an original estimate value, provide a value for the **Original Estimate** field when you are creating or editing an issue.

Logging work

Logging work in Jira allows you to specify the amount of time (work) you have spent working on an issue. You can log work against any of the issues, provided you have the permission to do so. We will cover permissions in `Chapter 10`, *Searching, Reporting, and Analysis*.

Go through the following steps to log work against an issue:

1. Browse to the issue you wish to log work against
2. Select **Log Work** from the **More** menu
3. Enter the amount of time you wish to log. Use w, d, h, and m to specify the week, day, hour, and minute, respectively
4. Select the date you wish to log your work against
5. Optionally, select how the remaining estimate should be adjusted
6. Add a description to the work you have done
7. Optionally, select who can view the work log entry
8. Click on the **Log** button

When you log work on an issue, you have the option to choose how the remaining estimate value will be affected. By default, this value will be automatically calculated by subtracting the amount that has been logged from the original estimate. You can, however, choose from other options that are available, such as setting the remaining estimate to a specific value or reducing it by an amount that is different from the amount of work being logged.

You can also click on the + sign in the **Time Tracking** section to log time.

Issues and comments

Jira lets users create comments on issues. As we have already seen, you will be able to create comments when assigning an issue to a different user. This is a very useful feature that allows multiple users to collaborate so that they can work on the same issue and share information. For example, the support staff (issue assignee) may request more clarification from the business user (issue reporter) by adding a comment to the issue. When combined with Jira's built-in notification system, automatic email notifications will be sent to the issue's reporter, assignee, and any of the other users watching the issue. Notifications will be covered in Chapter 8, *Emails and Notifications*.

Adding comments

By default, all logged-in users will be able to add comments to issues they can access. Go through the following steps to add a comment to an issue:

1. Browse to the issue you wish to add a comment to.
2. Click on the **Comment** button or press *M* on your keyboard.
3. Type a comment into the text box. You can preview and set restrictions on who can view your comment.
4. Click on the **Add** button to add the comment, as shown in the following screenshot:

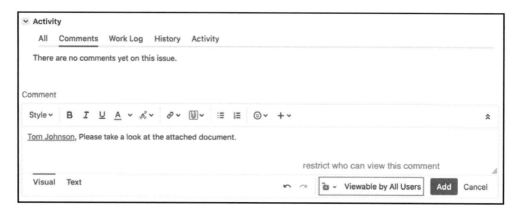

Once a comment has been added, the comment will be visible in the **Comments** tab in the **Activity** section at the bottom. When you are creating comments, you can select who can view your comment using the comment access control. This is very useful if you have external users viewing the issue and you only want to share your comments with internal users.

After you have added your comment to an issue, you can edit its contents and security settings or delete it altogether. To edit or delete a comment, simply hover over the comment and the comment management option will appear on the right-hand side, as shown in the following screenshot:

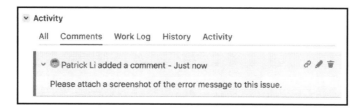

Permalinking a comment

From time to time, you will want to refer other people to a comment you have made previously. While you can tell them about the issue and let them scroll down to the bottom until they find your comment among hundreds of others, Jira allows you to create a quick permalink to your comment that will take you directly to the comment of interest.

Go through the following steps to create a permalink for a comment:

1. Browse to the comment you wish to create a permalink for
2. Hover over the comment to bring up the comment management options
3. Click on the permalink icon. This will highlight the comment in pale blue

You will now notice that your browser's URL bar will have something similar to `http://sample.jira.com/browse/DEMO-1?focusedCommentId=10100&page=com.atlassian.jira.plugin.system.issuetabpanels:comment-tabpanel#comment-10100` as a sample link (note the `focusedCommendId` phrase after the issue key). Copy and paste that URL and give it to your colleagues; once they click on this link, they will be taken directly to the highlighted comment.

Attachments

As we have seen so far, Jira uses fields such as **Summary** and **Description** to capture data. This works for most cases, but when you have complex data such as application log files or screenshots, this becomes insufficient. This is where attachments come in. Jira allows you to attach files from your local computer or a screenshot you have taken.

Attaching files

The easiest way to attach a file to a Jira issue is via the drag and drop action. You can do this by going through the following steps:

1. Browse to the issue you wish to attach a file to.

2. Drag and drop the files you want to attach to the browser. You will see an outline indicating where you can drop the file to attach it to the issue, as shown in the following screenshot:

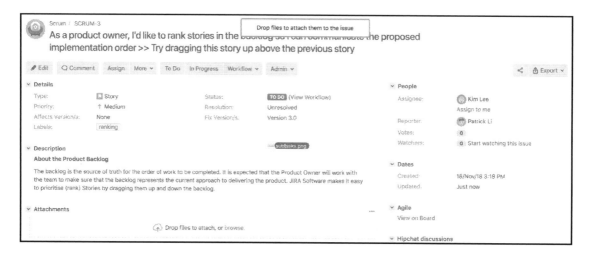

Drag and drop is the easiest way to attach files, but if, for some reason, drag and drop does not work, you can also manually select the file and attach it by going through the following steps:

1. Browse to the issue you wish to attach a file to
2. Select the **Attach files** option from the **More** menu
3. Select and attach the files you want to attach from the file browser

Depending on the file's type, certain files such as images and PDFs can be viewed directly from Jira's UI without having to download it.

Issue types and subtasks

As we saw earlier, issues in Jira can represent many things, ranging from software development tasks to project management milestones. Issue types are what differentiate one kind of issue from another.

Each issue has a type (hence the name issue type), which is represented by the issue type field. This lets you know what type of issue it is and helps you determine many other details, such as which fields will be displayed for this issue.

The default issue types are great for simple software development projects, but they do not necessarily meet the needs of others. Since it is impossible to create a system that can address everyone's needs, Jira lets you create your own issue types and assigns them to projects. For example, for a help desk project, you might want to create a custom issue type called **ticket**. You can create this custom issue type and assign it to the **Help Desk** project and users will be able to log tickets, instead of bugs, in the system.

Issue types are managed through the **Manage Issue Types** page. Perform the following steps to access this page:

1. Log in to Jira as a Jira administrator.
2. Browse to the Jira administration console.
3. Select the **Issues** tab and then the **Issue types** option. This will take you to the **Issue Types** page.

The following screenshot shows a list of default issue types that come with Jira Software. If you only have Jira Core, the list may look different:

Issue types				Add issue type ⑦
Name	**Type**	**Related Schemes**	**Actions**	
Bug A problem which impairs or prevents the functions of the product.	Standard	• SCRUM: Scrum Issue Type Scheme • KANBAN: Kanban Issue Type Scheme	Edit Delete Translate	
Epic Created by Jira Software - do not edit or delete. Issue type for a big user story that needs to be broken down.	Standard	• Default Issue Type Scheme • SCRUM: Scrum Issue Type Scheme • KANBAN: Kanban Issue Type Scheme	Edit Delete Translate	
Story Created by Jira Software - do not edit or delete. Issue type for a user story.	Standard	• Default Issue Type Scheme • SCRUM: Scrum Issue Type Scheme • KANBAN: Kanban Issue Type Scheme	Edit Delete Translate	
Task A task that needs to be done.	Standard	• TEST: Project Management Issue Type Scheme • SCRUM: Scrum Issue Type Scheme • PUB: Task Management Issue Type Scheme • PUB: Project Management Issue Type Scheme • PM: Project Management Issue Type Scheme • HR: Task Management Issue Type Scheme • KANBAN: Kanban Issue Type Scheme	Edit Delete Translate	

Creating issue types

You can create any number of issue types. Go through the following steps to create a new issue type:

1. Browse to the **Issue Types** page
2. Click on the **Add Issue Type** button
3. Enter the name and description for the new issue type
4. Select whether the new issue type will be a standard issue type or a subtask issue type
5. Click on **Add** to create the new issue type

Once the new issue type has been created, it will be assigned a default icon. If you want to change the icon, you will need to click on the **Edit** link for the issue type and then select a new image as its icon.

Deleting issue types

When deleting an issue type, you have to keep in mind that the issue type might already be in use, meaning that issues have already been created with that issue type. So, when you delete an issue type, you will need to select a new one for those issues. The good news is that Jira takes care of this for you. As we can see in the following screenshot, when we delete the **Bug** issue type, Jira informs us of the already existing 13 issues that are of the **Bug** type. You will need to assign them to a new issue type, such as **Improvement**:

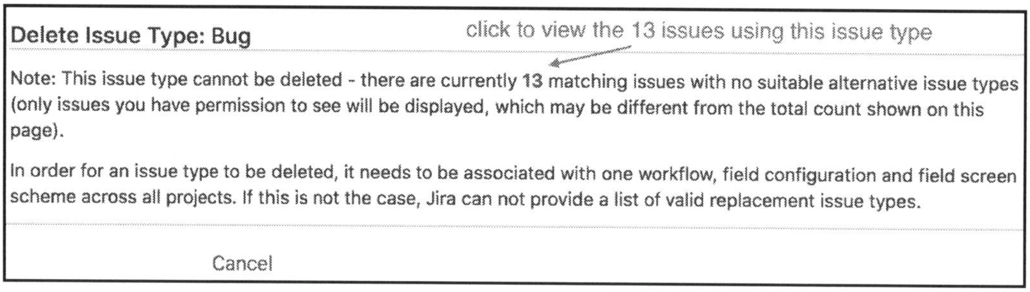

Delete Issue Type: Bug   click to view the 13 issues using this issue type

Note: This issue type cannot be deleted - there are currently **13** matching issues with no suitable alternative issue types (only issues you have permission to see will be displayed, which may be different from the total count shown on this page).

In order for an issue type to be deleted, it needs to be associated with one workflow, field configuration and field screen scheme across all projects. If this is not the case, Jira can not provide a list of valid replacement issue types.

Cancel

Subtasks

Jira allows only one person (assignee) to work on one issue at a time. This design ensures that an issue is a single unit of work that can be tracked against one person. However, in the real world, we often find ourselves in situations where we need to have multiple people working on the same issue. This may be caused by a poor breakdown of tasks or simply because of the nature of the task at hand. Whatever the reason, Jira provides a mechanism to address this problem through subtasks.

Subtasks are similar to issues in many ways, and as a matter of fact, they are a special kind of issue. They must have a parent issue, and their issue types are flagged as subtask issue types. You can say that all subtasks are issues, but not all issues are subtasks.

For every issue, you can have one or more subtasks that can be assigned and tracked separately from each other. Subtasks cannot have other subtasks. Jira only allows one level of subtask.

Creating subtasks

Since subtasks belong to an issue, you need to browse to the issue first before you can create a new subtask by going through the following steps.:

1. Browse to the issue you wish to create subtasks for
2. Select **Create Sub-Task** from the **More** menu

You will see a familiar **Create Issue** dialog box; however, one thing you will notice is that, unlike when you are creating an issue, you do not select which project to create the subtask in. This is because Jira can determine the project's value based on the parent issue. You will also notice that you can only select issue types that are subtasks.

Other than these differences, creating a subtask is no different than creating a normal issue. You can customize the fields that are shown in the dialog box and choose to rapidly create multiple subtasks by selecting the **Create another** option.

Once the subtask has been created, it will be added to the **Sub-Tasks** section of the parent issue. You will see all the subtasks that belong to the issue and their status. If a subtask has been completed, it will have a green tick next to it, as shown in the following screenshot:

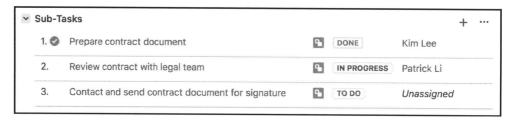

Issue type schemes

Issue type schemes are templates or collections of issue types that can be applied to projects. As we can see in the following screenshot, Jira comes with a default issue type scheme, which is applied to all projects that do not have specific issue type schemes applied. When you create a new project, a new issue type scheme is created for you based on the project template you have selected. The new scheme will also have issue types pre-populated based on the template. As we can see in the following screenshot, we have two issue type schemes, TEST: Project Management Issue Type Scheme for test project and SCRUM: Scrum Issue Type Scheme not being used:

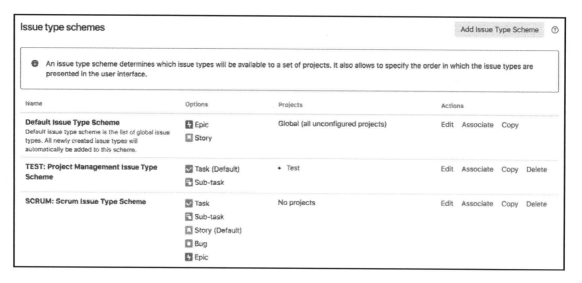

When you create your own issue types, to make them available, you need to add them to the issue type scheme that's used by your project.

Adding issue types to an issue type scheme

Go through the following steps to create a new issue type scheme:

1. Browse to the administration console.
2. Select the **Issues** tab and then the **Issue type schemes** option. This will bring you to the **Issue Type Schemes** page.
3. Click on the **Edit** link for the issue type scheme you want to add issue types to.
4. Drag the issue types that you want to be part of the scheme from the **Available Issue Types** list and drop them into the **Issue Types for Current Scheme** list.
5. Select a **Default Issue Type** value. Note that this is optional, and you can only select a default issue type after you have selected at least one issue type for the new scheme.
6. Click on the **Save** button, as shown in the following screenshot:

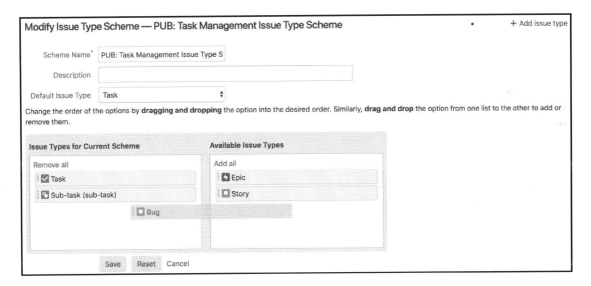

Issue priorities

Priorities help users set the importance of issues. Users can first assign priority values to issues and later use them to sort the list of issues they have to work on, thereby helping the team decide which issues to focus on first. Jira comes with five levels of priorities out of the box, as shown in the following screenshot:

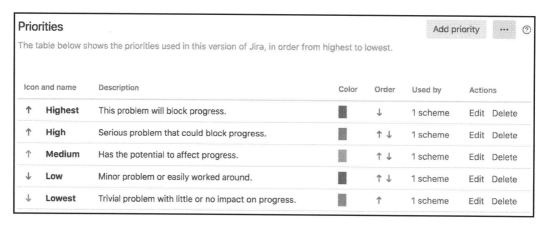

You can customize this list by creating your own priorities. To create new priorities, follow these steps:

1. Browse to the administration console.
2. Select the **Issues** tab and then the **Priorities** option.
3. Enter a name and description for the new priority.
4. Click on the **select image** link to choose an icon for the priority.
5. Specify a color for the priority. You can either type in the HTML color hex code directly or use the color picker to help you select the color you want. The color that's chosen here will be used when icon images cannot be displayed, such as when you export issues to a spreadsheet.
6. Click on the **Add** button.

Creating a priority scheme

Issue priorities are used globally in Jira, so all projects will have the same set of priority options, although this is a limitations that has caused inconveniences when certain projects need to have their own set of priority options. Atlassian has addressed this in recent Jira updates to the software so that the priority scheme feature is included.

Priority schemes work in a similar way to the issue type scheme feature we looked at earlier. You can create a scheme so that it contains only the issue priorities you need and then apply the scheme to a project. This way, each project can have its own set of priority options. To create and apply a new priority scheme, follow these steps:

1. Browse to the administration console.
2. Select the **Issues** tab and then the **Priority schemes** option.
3. Click on the **Add priority scheme** button.
4. Enter a name for the new priority scheme.
5. Drag the priorities that you want to be part of the scheme from the **Available priorities** list and drop them into the **Selected priorities** list.
6. Select a **Default priority** value. Note that this is optional, and you can only select a default priority after you have selected at least one priority for the new scheme.
7. Click on the **Add** button, as shown in the following screenshot:

Once you have created the new priority scheme, you can go to your project's setting page, select the new priority scheme, and apply the scheme.

The HR project

In this exercise, we will continue our setup for the project we created in the previous chapter. We will add the following configurations to our project:

- A set of new issue types that are specific to our HR project
- Add our new issue types to the issue type scheme to make them available

Adding new issue types

Since our project is for the human resources team, we need to create a few custom issue types to augment the default ones that come with Jira. For this exercise, we will create two new issue types, New Employee and Termination.

The first step, that is, setting up an issue type association, is to create the two issue types that we need, New Employee and Termination:

1. Browse to the **Issue Types** page
2. Click on the **Add Issue Type** button
3. Type New Employee in the **Name** field
4. Click on the **Add** button to create the new issue type

You should now see the new issue type in the table. Now, let's add the Termination issue type:

1. Click on the **Add Issue Type** button again
2. Type Termination in the **Name** field
3. Click on **Add** to create the new issue type

You should see both the New Employee and Termination issue types. However, this will only make our new issue types available—it will not make them the only options when creating a new issue for our project. As you may recall from the previous sections, we need to add the new issue types to the issue type scheme that's used by our project.

Updating the issue type scheme

We want to limit the issue types to be only `New Employee`, `Termination`, and the generic `Task` for our `HR` project, but we do not want to affect the other projects that still need to have **Bug** and other default issue types. Therefore, we need to create a new issue type scheme specifically for our project. We can do this by going through the following steps:

1. Browse to the **Issue Type Schemes** page.
2. Click on the **edit** link for our issue type scheme. The default one that's created by Jira should be called `HR: Task Management Issue Type Scheme`.
3. Drag the `New Employee` and `Termination` issue types from the **Available Issue Types** panel to the **Issue Types for Current Scheme** panel.
4. Click on the **Save** button.

Putting it all together

With everything created and set up, you can go back and create a new issue to see how it all looks. If everything works out, you should see something similar to the following screenshot:

Summary

In this chapter, we looked at what issues are in Jira and explored the basic operations of creating, editing, and deleting issues. We also looked at the advanced operations offered by Jira to enhance how you can manipulate and use issues, such as adding attachments, creating subtasks, and linking multiple issues.

In the next chapter, we will look at fields and how we can create our own custom fields to capture additional information from users.

5
Field Management

Projects are collections of issues, and issues are collections of fields. As we have seen in the earlier chapters, fields capture data, which can then be displayed to users. There are many different types of field in Jira, ranging from simple text fields that let you input alphanumeric text, to more complicated fields with pickers to assist you with choosing dates and users.

An information system is only as useful as the data that goes into it. By understanding how to effectively use fields, you can turn Jira into a powerful information system for data collection, processing, and reporting.

In this chapter, we will expand our HR project with these customized fields and configurations, by exploring fields in detail and learning how they relate to other aspects of Jira. By the end of this chapter, you will have learned the following:

- Using system and custom fields
- Collecting custom data through custom fields
- Adding behaviors to fields with field configurations
- Understanding field configuration schemes and how to apply them to projects

Understanding system fields

Jira comes with a number of built-in fields. You have already seen a few of them in the previous chapters. Fields such as summary, priority, and assignee are all built-in. They make up the backbone of an issue, and you cannot remove them from the system. For this reason, they are referred to as **system fields**.

The following table lists the most important system fields in Jira:

System field	Description
Assignee	This specifies the user who is currently assigned to work on the issue.
Summary	This specifies a one-line summary of the issue.
Description	This provides a detailed description of the issue.
Reporter	This specifies the user who has reported this issue (the majority of the time, it is also the person who has created the issue, but not always).
Component/s	This specifies the project components the issue belongs to.
Effects Version/s	This specifies the versions the issue effects are found in.
Fix Version/s	This specifies the versions the issue will be fixed in.
Due Date	This specifies the date this issue is due.
Issue Type	This specifies the type of issue (for example, Bug and New Feature).
Priority	This specifies how important the issue is compared to other issues.
Resolution	This specifies the current resolution value of the issue (for example, Unresolved or Fixed).
Time Tracking	This lets users estimate how long the issue will take to be complete.

Understanding custom fields

While Jira's built-in fields are quite comprehensive for basic general uses, most organizations soon find they have special requirements that cannot be addressed simply with the system fields available. To help you tailor Jira to your organization's requirements, Jira lets you create and add your own fields to the system, called **custom fields**.

Every custom field belongs to a custom type, which dictates its behavior, appearance, and functionality. Therefore, when you add a custom field to Jira, you really add another instance of a custom field type.

Jira comes with over 20 custom field types that you can use straight out-of-the-box. Many custom field types are identical to the built-in fields, such as date picker, which is like the due date field. They provide you with simplicity and flexibility that are not available with their built-in counterparts. The upcoming tables break down and list all the standard Jira custom field types and their characteristics.

Standard fields

These fields are the most basic field types in Jira. They are usually simple and straightforward to use, such as the text field, which allows users to input any text:

Custom field type	Description
Date Picker	These are input fields that allow input with a date picker and enforce valid dates.
Date Time Picker	These are input fields that allow input with a date and time picker and enforce valid date timestamps.
Labels	These are input fields that allow tags to be added to an issue.
Number Field	These are input fields that store and validate numeric values.
Radio Buttons	These are radio buttons that ensure only one value can be selected.
Select List (cascading)	These are multiple select lists where the options for the second select list are dynamically updated based on the value of the first.
Select List (multiple choice)	These are multiple select lists with a configurable list of options.
Select List (single choice)	These are single select lists with a configurable list of options.
Text Field (multi-line)	These are multiple line text areas enabling the incorporation of significant text content.
Text Field (single-line)	These are basic single link input fields that allow simple text inputs of fewer than 255 characters.
URL Field	These are input fields that validate a valid URL.
User Picker (single user)	These choose a user from the Jira user base through either a pop-up user picker window or auto completion.

Advanced fields

These fields provide specialized functions. For example, the Version Picker field lets you select a version from the current project. If you have any custom fields from third-party add-ons (see the following section), they will also be listed here:

Custom field type	Description
Group Picker (multiple group)	This chooses one or more user groups using a pop-up picker window.
Group Picker (single group)	This chooses a user group using a pop-up picker window.
Project Picker (single project)	This selects lists displaying the projects that are viewable to the user in the system.
Text Field (read only)	This is a read-only text field that does not allow users to set their data. It's only possible to set the data programmatically.

User Picker (multiple users)	This chooses one or more users from the user base through a pop-up picker window.
Version Picker (multiple versions)	This chooses one or more versions from the available versions in the current project.
Version Picker (single version)	This chooses a single version from the available versions in the project.

As you can see, Jira provides you with a comprehensive list of custom field types. In addition, there are many custom field types developed by third-party vendors (available as **plugins** or **add-ons**) that you can add to your Jira to enhance its functionality. These custom fields provide many specialized functionalities, such as automatically calculating values and retrieving data from databases directly or connecting to an external system. Once you install the plugin, the process of adding custom fields from other vendors is mostly the same as adding custom fields shipped with Jira.

The following list shows some examples of add-ons that provide additional useful custom fields. You can find them from the Atlassian Marketplace at `https://marketplace.atlassian.com`:

- **Enhancer Plugin for Jira**: This includes a number of custom fields that will automatically display dates when key events occur for an issue; for example, when the issue was last closed
- **Toolkit Plugin for Jira**: This provides a number of useful custom fields, such as showing statistics on users that participate in a given issue and the date when the issue was last commented on
- **nFeed**: This provides a suite of custom fields that let you connect to databases, remote files, and web services to retrieve data and display it in Jira
- **21 CFR Part 11 E-Signature**: This lets users electronically sign issues in Jira as they work on them, for example, approving an issue to be closed
- **SuggestiMate for Jira**: This provides a specialized custom field that shows similar and potentially duplicated issues when creating new issues or browsing through existing ones

Understanding field searchers

For any information system, capturing data is only half of the equation. Users will need to be able to retrieve the data at a later stage, usually through searching, and Jira is no different. While fields in Jira are responsible for capturing and displaying data, it is their corresponding searchers that provide the search functionality.

All fields that come with Jira have searchers associated with them by default, so you will be able to search issues according to their summary or assignee, without any further configuration. Some custom fields from third-party add-ons may have more than one searcher available. You can change the default searcher by editing the custom field.

In the Jira UI, a searcher is referred to as a search template.

Custom field context

System fields, such as resolution, are global across Jira. What this means is that these fields will have the same set of selections for all projects. Custom fields, on the other hand, are a lot more flexible.

Custom field types, such as select lists and radio buttons, can have different sets of options for different projects or different issue types within the same project. This is achieved through what is called a **custom field context**.

A custom field context is made up of a combination of projects and issue types. When you are working with an issue, Jira will check the project and issue type of the current issue to determine whether there is a specific context that matches the combination. If one is found, Jira will load the custom field with any specific settings, such as selection options. However, if no context is found, the custom field will not be loaded.

In Jira, if no context can be found that matches the project and issue type combination, a custom field does not exist for the issue.

We will look at how to set custom field contexts in a later section. What you need to remember now is that when adding a custom field, you need to make sure that it has the correct context setting.

Managing custom fields

Custom fields are used globally across Jira, so you will need to have the Jira Administrator global permission to carry out management operations such as creation and configuration.

Jira maintains all the custom fields in a centralized location for easy management. Perform the following steps to access the custom field management page:

1. Log in as a Jira administrator user.
2. Browse to the Jira administration console.
3. Select the **Issues** tab and then the **Custom fields** option:

On the **Custom fields** page, all the existing custom fields will be listed. From here, you can see the names of all custom fields, their type, the context they belong to, and the screens they are displayed on. Note that some custom fields, such as Development and Epic Color, as shown in the previous screenshot, come with Jira itself, and will have the **LOCKED** label next to their names. These fields serve special purposes in Jira, so their configurations cannot be changed. User-added custom fields, such as Department, do not have this restriction and can be updated at any time.

Adding a custom field

Adding a new custom field is a multistep process, and Jira provides a wizard to help you through it. There are two mandatory steps and an optional step when adding a new custom field. You need to first select the type of custom field, then its name, followed by options if you are adding a select list custom field type. The final, optional, step is to decide which screens to add the field onto. We will walk you through the process:

1. Browse to the **Custom fields** page.
2. Click on the **Add custom field** button. This will bring you to step 1 of the process, where you can select the custom field type.

3. Search and select the custom field type you wish to add, and click on **Next**. This will bring you to step 2 of the process, where you can specify the custom field's name and options:

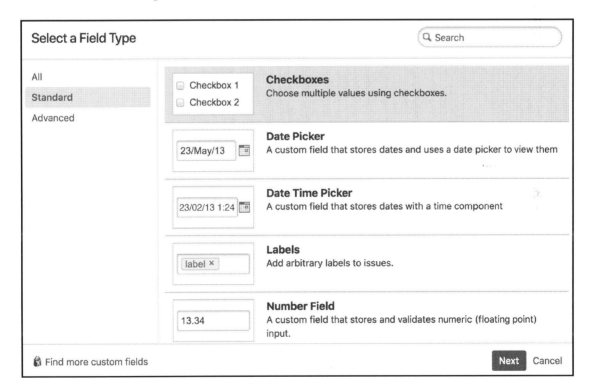

If you do not see the field type you are looking for, select the **All** option from the left-hand side and then search again.

4. Enter values for the **Name** and **Description** fields. If you are creating a selection-based custom field, such as a select list, you will need to add its select options, too (you can update this list later):

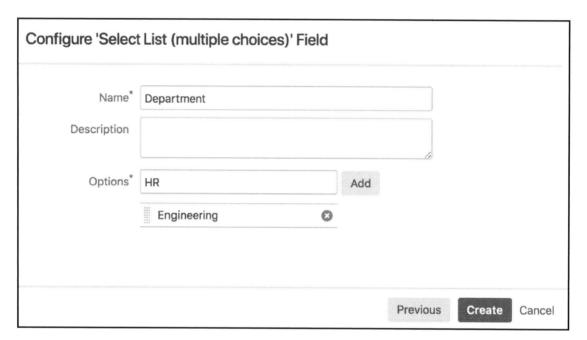

Even though you can have multiple custom fields with the same name, this is usually not a good practice as it can lead to confusion later on and make management difficult.

5. Click on the **Create** button. This will bring you to the final step of the process, where you can specify which screen you would like to add the field onto. This step is optional, as the custom field has already been added in Jira. You do not have to add the field onto a screen. We will discuss fields and screens in Chapter 6, *Screen Management*.

6. Select the screens and click on **Update**. The following screenshot shows that the newly created field has been added to **Default Screen**:

Associate field Department to screens

Associate the field Department to the appropriate screens. You must associate a field to a screen before it will be displayed. New fields will be added to the end of a tab.

Screen	Tab	Select
Default Screen	Field Tab	☑
HR: Task Management Create Issue Screen	Field Tab	☐
HR: Task Management Edit/View Issue Screen	Field Tab	☐
Resolve Issue Screen	Field Tab	☐
Workflow Screen	Field Tab	☐

Update   Cancel

Once a custom field has been created, you will see it on the appropriate screen when you are creating, editing, or viewing issues.

Editing/deleting a custom field

Once a custom field has been created, you can edit its details at any time. You may already notice that there is a **Configure** option and an **Edit** option for each custom field. It may be confusing in the beginning to differentiate between the two. **Configure** specifies options related to the custom field context, which we will discuss in the following sections. Edit specifies options that are global across Jira for the custom field; these include its name, description, and search templates:

1. Browse to the **Custom fields** page
2. Select the **Edit** option by clicking on the cog icon for the custom field you wish to edit from the list of custom fields
3. Change the custom field's details, such as its name or search template
4. Click on the **Update** button to apply the changes

When making changes to the search templates for your custom fields, it is important to note that, while the change will take effect immediately, you need to perform a full system re-index in order for Jira to return the correct search results. This is because, for each search template, the underlying search data structure may be different, and Jira will need to update its search index for the newly applied search template.

For example, if you have a custom field that did not have a searcher and you have just applied a searcher to it, no results will be returned until you re-index Jira. When you make changes to the search template, Jira will alert you with a message that a re-index will be required, as shown in the following screenshot:

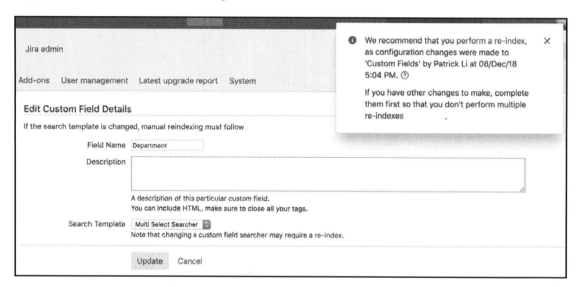

 You should select the background re-index option to avoid any downtime.

We will discuss searching and indexing in more detail in `Chapter 10`, *Searching, Reporting, and Analysis*.

You can also delete existing custom fields, as follows:

1. Browse to the **Custom fields** page
2. Select the **Delete** option by clicking on the tools icon for the custom field you wish to delete
3. Click on the **Delete** button to delete the custom field

Once deleted, you cannot get the custom field back, and you will not be able to retrieve and search the data held by those fields. If you try to create another custom field of the same type and name, it will not inherit the data from the previous custom field, as Jira assigns unique identifiers to each of them. It is highly recommended to back up your Jira project before you delete the field unless you are absolutely sure you do not need it.

Configuring a custom field

Now that we have seen how to create and manage custom fields, we can start looking at more advanced configuration options. Different custom field types will have different configuration options available to them. For example, while all custom fields will have the option to specify one or more contexts, selection list-based custom fields will also allow you to specify a list of options. We will look at each of the configuration options in the following sections.

To configure a custom field, you need to access the **Configure Custom Field** page, as follows:

1. Browse to the **Custom fields** page.
2. Select the **Configure** option by clicking on the cog icon for the custom field you wish to configure from the list of custom fields. This will bring you to the **Configure Custom Field** page.

The following screenshot shows that the **Source** custom field has two available contexts, the default context **Default Configuration Scheme**, which is applied to all projects except **Project Management**(this has its own context) and **PMO Configuration Scheme**, which is applied only to the **Project Management** project:

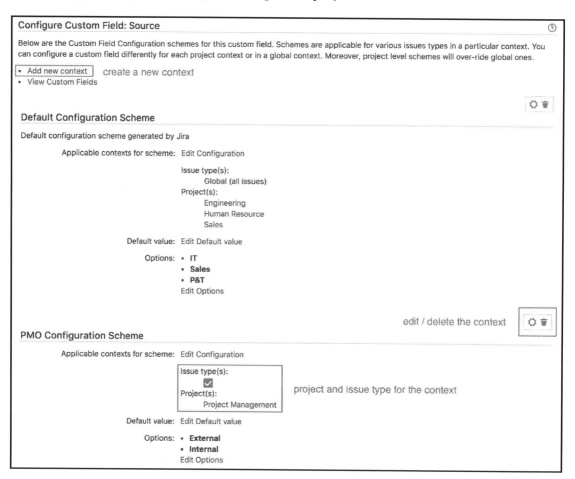

Adding custom field contexts

From time to time, you may need your custom fields to have different behaviors, depending on what project the issue is located in. For example, if we have a select list custom field called `Department`, we may want it to have a different set of options based on which project the issue is being created in, or even a different default value.

To achieve this level of customization, Jira allows you to create multiple custom field contexts for a custom field. As we have seen already, a custom field context is a combination of issue types and projects. Therefore, in our preceding example, we have a context for the `Task` issue type and the `Project Management` project with a different set of options.

Jira allows you to configure custom fields based on issue types and projects through contexts. Each project can have only one configuration context per custom field.

Creating a new custom field context is simple. All you need to do is decide the issue type and project combination that will define the context:

1. Browse to the **Configure Custom Field** page for the custom field you wish to create a new context for.
2. Click on the **Add new context** link. This will take you to the **Add configuration scheme context** page.
3. Enter a name in the new custom field context in the **Configuration scheme** label field.
4. Select the issue types for the new context under the **Choose applicable issue types** section.
5. Select projects for the new context under the **Choose applicable context** section.
6. Click on the **Add** button to create the new custom field context.

Each project can only belong to one custom field context per custom field (global context is not counted for this). Once you select a project for context, it will not be available the next time you create a new context. For example, if you create a new context for `Project A`, it will not be listed as an option when you create another context for the same custom field. This is to prevent you from accidentally creating two contexts for the same project.

After a new custom field context has been created, it will not inherit any configuration values as the default context, such as the **Default Value** and **Select Options** from other contexts. You will need to repopulate and maintain the configuration options for each newly created context.

Restricting custom field's context can also help reduce search index size and help improve performance, especially with data center edition deployment.

Configuring select options

For custom field types, select a list, checkboxes, radio buttons, and their multi-versions. You need to configure their select options before they can become useful to users. Select options are configured and set on a per-custom-field-context basis. This provides the custom field with the flexibility of having different select options for different projects.

To configure select options, you need to first select the custom field and then the context that the options will be applied to, as follows:

1. Browse to the **Custom fields** page.
2. Click on the **Configure** option for the custom field you wish to configure the select options for.
3. Click on the **Edit Options** link for the custom field context you wish to apply the options.
4. Enter option values in the **Add New Custom Field Option** section, and click on the **Add** button to add the value. The options will be added in the order in which they are entered into the system. You can manually move option values up and down or click on **Sort options alphabetically** to let Jira perform the sorting for you.
5. Click on the **Done** button once you finish configuring the select options:

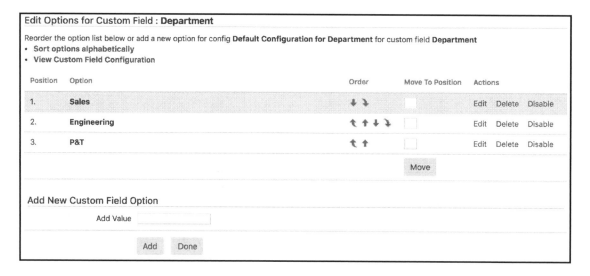

Setting default values

For most custom fields, you can set a default value so your users will not need to fill them in unless they have special needs. For text-based custom fields, the default values will be displayed as text by default, when the users create or edit an issue. For selection-based custom fields, the default values will be pre-selected options for users.

Just like setting selection options, default options are also set on a per-custom-field-context basis:

1. Browse to the **Custom fields** page
2. Click on the **Configure** option for the custom field for which you wish to configure select options
3. Click on the **Edit Default Value** link for the custom field context to which you want to apply the default values
4. Set the default value for the custom field
5. Click on the **Set Default** button to set the default value

Setting the default value will be different for different custom field types. For text-based custom fields, you will be able to type any text string. For select-based custom fields, you will be able to select from the options you add. For picker-based custom fields, such as User Picker, you will be able to select a user directly from the user base.

Field configuration

As you have already seen, fields are used to capture and display data in Jira. Fields can also have behaviors, which are defined by field configuration. For each field in Jira, you can configure its behaviors, which are listed as follows:

- **Field description**: This is the description text that appears under the field when an issue is edited. With field configuration, you can have different description text for different projects and issue types.
- **Visibility**: This determines whether a field should be visible or hidden.
- **Required**: This specifies whether a field will be optional or required to have a value when an issue is being created/updated. When applied to select, checkbox, or radio button custom fields, this will remove the **None** option from the list.
- **Rendering**: This specifies how the content is to be rendered for text-based fields (for example, a wiki renderer or a simple text renderer for text fields).

A field configuration provides you with control over each individual field in your Jira, including both built-in and custom fields. Since it is usually a good practice to reuse the same set of fields instead of creating new ones for every project, Jira allows you to create multiple field configurations, by means of which we can specify different behaviors on the same set of fields and apply them to different projects.

We will be looking at how to manage and apply multiple field configurations in the later sections of this chapter. But first, let's take a close look at how to create new field configurations and what we can do with them.

You can access the field configuration management page through the Jira administration console:

1. Browse to the Jira administration console.
2. Select the **Issues** tab and then the **Field configurations** option. This will bring you to the **View Field Configurations** page.

Adding a field configuration

Creating new field configurations is simple. All you need to do is specify the name and a short description for the new configuration:

1. Browse to the **View Field Configurations** page
2. Click on the **Add Field Configuration** button
3. Enter the name and description for the new field configuration
4. Click on the **Add** button to create a field configuration

As we will see later in the *Field Configuration Scheme* section, field configurations are linked to issue types, so it is recommended to name them based on the issue type they will be applied to and with a version number at the end, for example, Bugs Field Configuration 1.0. This way, when you need to make changes to the field configuration, you can increment the version number, leaving a history of changes you can revert to.

After a field configuration is created, it is not put to use until we associate it with a field configuration scheme. We will look at how to do this in later sections.

Managing field configurations

Now that we have seen how to create new field configurations, it is time for us to take a closer look at the different configuration options. Firstly, just a quick recap—each field configuration includes all the fields available in Jira, and its behavior is defined relevant to each field configuration. We will then associate it with a field configuration scheme, which will determine when a field configuration will become active for a given issue.

Perform the following steps to access field configuration options:

1. Browse to the **View Field Configurations** page.
2. Click on the **Configure** link for the field configuration you wish to configure. This will take you to the **View Field Configuration** page.

On this page, all the fields and their current configuration options that are currently set for the selected field configuration are listed:

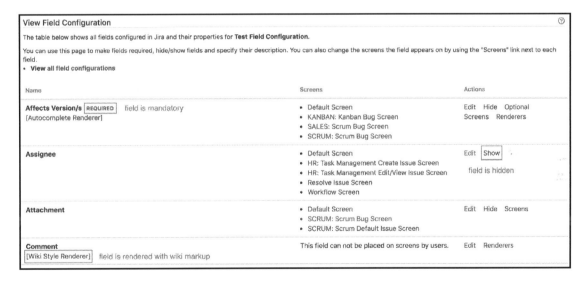

As you can see, there are several options you can configure for each field, and depending on the field type, the options may vary. While we will be looking at each of the options, it is important to note that some options will override each other. This is Jira trying to protect you from accidentally creating a configuration combination that will break your system. For example, if a field is set to both hidden and required, your users will not be able to create or edit issues, so Jira will not allow you to set a field to **REQUIRED** if you have already set it to hidden.

Field description

While having a meaningful name for your fields will help your users understand what the fields are for, providing a short description will provide more context and meaning. Field descriptions are displayed under the fields when you create or edit an issue. To add a description for a field, do the following:

1. Browse to the **View Field Configuration** page for the field configuration you wish to use
2. Click on the **Edit** link for the field for which you wish to set a description
3. Add a description for the field, and click on **Update**

 For custom fields, the description you enter here will override the description you provide when you first create them.

Field requirement

You can set certain fields as required or mandatory for certain issues. This is a very useful feature as it ensures that critical information can be captured when users create issues. For example, for our support system, it makes sense to have our users enter in the system that is misbehaving in a field and make that field compulsory to help our support engineers.

You have already seen required fields in action. System fields, such as `Summary` and `Issue Type`, are compulsory in Jira (and you cannot change that). When you do not specify a value for a required field, Jira will display an error message underneath the field, telling you that the value is required.

When you add a new field into Jira, such as custom fields, it is optional by default, meaning users do not need to specify a value. You can then change the setting to make those fields required:

1. Browse to the **View Field Configuration** page for the field configuration you wish to use
2. Click on the **Required/Optional** link for the field you wish to set as the mandatory requirement

You will notice that once a field is set to required, there will be a small required text label in red next to the field name. When you create or edit an issue, the field will have a red (*) character next to its name. This is Jira's way of indicating that a field is mandatory.

Field visibility

Most fields in Jira can be hidden from a user's view. When a field is set to hidden, users will not see the fields on any screens, including issues such as create, update, and view. Perform the following steps in order to show or hide a field:

1. Browse to the **View Field Configuration** page for the field configuration you wish to use
2. Click on the **Show/Hide** link for the field you wish to show or hide, respectively

Once a field has been set to hidden, it will not appear on screen and you will not be able to search in it. However, you can still use tools such as scripts to set values for hidden fields. For this reason, hidden fields are used to store data that is used by automated processes.

Not all fields can be hidden. Built-in fields, such as Summary and Issue Type, cannot be hidden. When you set a field to hidden, you will notice that you can no longer set the field as required. As stated earlier, setting a field to required will make Jira enforce a value to be entered into the field when you create or edit an issue. If the field is hidden, there will be no way for you to set a value and you will be stuck. This is why Jira will automatically disable the required option, especially if you have already hidden a field. On the other hand, if you marked a field as required, when you hide the same field you will notice that the field is no longer required. The rule of thumb is that field visibility will override field requirement.

 A field cannot be both hidden and required.

Field rendering

Renderers control how a field will be displayed when it is being viewed or edited. Some built-in and custom fields have more than one renderer, and for these fields, you can choose which one to use. For example, for text-based fields, such as Description, you can choose to use the simple text renderer or the more sophisticated wiki-style renderer that will allow you to use wiki markup to add more styling.

Jira ships with four different renderers:

- **Default text renderer**: This is the default renderer for text-based fields. Contents are rendered as plain text. If the text resolves a Jira issue key, the renderer will automatically turn that into an HTML link.

- **Wiki style renderer**: This is an enhanced renderer for text-based fields. It allows you to use wiki markup to decorate your text content.
- **Select list renderer**: This is the default renderer for selection-based fields. It is rendered as a standard HTML select list.
- **Autocomplete renderer**: This is an enhanced renderer for selection-based fields, and it provides an autocomplete feature to assist users as they start typing into the fields.

The following table lists all the fields that can have special renders configured and their available options:

Field	Available renderers
Description	This is a wiki-style renderer and default text renderer.
Comment	This is a wiki-style renderer and default text renderer.
Environment	This is a wiki-style renderer and default text renderer.
Component	This is an autocomplete renderer and select list renderer.
Affects version	This is an autocomplete renderer and select list renderer.
Fix versions	This is an autocomplete renderer and select list renderer.
Custom field of type `Free Text Field` (unlimited text)	This is a wiki-style renderer and default text renderer.
Custom field of type `Text Field`	This is a wiki-style renderer and default text renderer.
Custom field of type `Multi Select`	This is an autocomplete renderer and select list renderer.
Custom field of type `Version Picker`	This is an autocomplete renderer and select list renderer.

Perform the following steps to set the renderer for a field:

1. Browse the **View Field Configuration** page for the field configuration you wish to use.
2. Click on the **Renderer** link for the field you wish to set a renderer for (if it is available). You will be taken to the **Edit Field Renderer** page.
3. Select the renderer from the available drop-down list.
4. Click on the **Update** button to set the renderer.

There are other custom renderers developed by third-party vendors. Just like custom fields, these are packaged as add-ons that you can install in Jira. Once installed, these custom renderers will be available for the selection of the appropriate field types.

A good example is the **JEditor** add-on, which provides a rich-text editor for all text-based fields including `Description`.

Screens

In order for a field to be displayed when you view, create, or edit an issue, it needs to be placed on a screen. You have already seen this when creating new custom fields. One of the steps in the creation process is to select what screens to add the custom field to. Screens will be discussed further in Chapter 6, *Screen Management*, so we will not spend too much time understanding them right now.

What you need to know for now is that after a field has been added to a screen, you can add it to additional screens or take it off completely. If you are working with just one field, you can configure it here from the field configurations. If you have multiple fields to update, a better approach will be to work directly with screens, as we will see in Chapter 6, *Screen Management*.

There is a subtle difference between hiding a field in field configuration and not placing a field on a screen. While the end result will be similar where, in both cases, the field will not show up, if you hide a field, you can still set a value for it through the use of default value, workflow post-functions (covered in Chapter 7, *Business Process and Workflow*), or custom scripts, essentially meaning that the field is there but just hidden. However, if the field is not on the screen, you cannot set its value, and so it can be considered as not being part of the issue. Another difference is that hiding a field will hide it for all screens that have the field added, for projects using the field configuration.

Field configuration scheme

With multiple field configurations, Jira determines when to apply each of the configurations through the field configuration scheme. A field configuration scheme maps field configurations to issue types. This scheme can then be associated with one or more projects.

This allows you to group multiple field configurations mapped to issue types and apply them to a project in one go. The project will then be able to determine which field configuration to apply, based on the nature of the issue. For example, for a given project, you can have different field configurations for bugs and tasks.

This grouping of configurations into schemes also provides you with the option to reuse existing configurations without duplicating work, as each scheme can be reused and associated with multiple projects.

Managing field configuration schemes

You can manage all your field configuration schemes from the **View Field Configuration Schemes** page. From there, you will be able to add, configure, edit, delete, and copy schemes:

1. Browse to the Jira administration console.
2. Select the **Issues** tab and then the **Field Configuration Schemes** option. This will bring you to the **View Field Configuration Schemes** page:

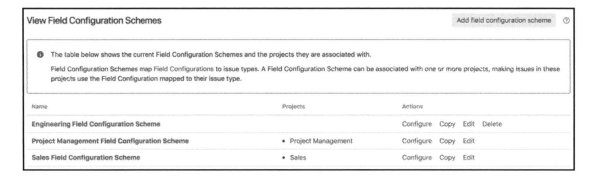

Adding a field configuration scheme

The first step in grouping your field configurations is to create a new field configuration scheme. By default, Jira does not come with any field configuration schemes. All the projects will use the system default field configuration. The new field configuration scheme will hold all the mappings between our field configurations and issue types.

To create a new field configuration scheme, all you need to do is specify the name and an optional description for the scheme:

1. Browse to the **View Field Configuration Schemes** page
2. Click on the **Add Field Configuration Scheme** button
3. Enter a name and description for the new field configuration scheme
4. Click on the **Add** button to create the scheme

Since field configuration schemes are applied to projects, it is good practice to name them according to the projects. For example, the scheme for the sales project can be named `Sales Field Configuration Scheme`. You can add a version number after the name to help you maintain changes.

Once the new field configuration scheme is created, it will be displayed in the table that lists all the existing schemes. At this time, the scheme is not yet useful as it does not contain any configuration mappings and is associated with a project.

Configuring a field configuration scheme

Once you have a new field configuration scheme set up, you will be able to add mapping between field configurations and issue types. For each field configuration scheme, one issue type can be mapped to only one field configuration, while each field configuration can be mapped to multiple issue types. The following screenshot shows that the issue types, **Epic**, **Story**, and **Task**, all have specific field configurations applied, and that the **Default Field Configuration** will be applied to all other issue types that are not explicitly mapped, such as Bug:

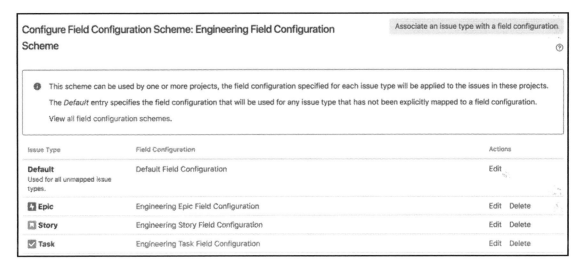

 One issue type can only be mapped to one field configuration.

When a field configuration scheme is first created, Jira creates a default mapping, which maps all unmapped issue types to the default field configuration. You cannot delete this default mapping as it acts as a catch-all condition for mappings that you do not specify in your scheme. What you need to do is add more specific mappings that will take precedence over this default mapping:

1. Browse to the **View Field Configuration Schemes** page
2. Click on the **Configure** link for the field configuration scheme you wish to configure
3. Click on the **Associate an Issue Type with a Field Configuration** button
4. Select the issue type and field configuration from the dialog
5. Click on the **Add** button to add the mapping

You can repeat these steps to add more mapping for other issue types. All unmapped issue types will use the **Default** mapping.

Associating a field configuration scheme with a project

After you create a new field configuration scheme and establish the mappings, the final step is to associate the scheme with a project for the configurations to take effect.

It is important to note that, once you associate the field configuration scheme with a project, you cannot delete it until you remove all the associations so that the scheme becomes inactive again.

To associate a field configuration scheme with a project:

1. Browse to the target project's administration page
2. Click on the **Fields** option in the left panel
3. Select the **Use a different scheme** option from the **Actions** menu
4. Select a new field configuration scheme and click on the **Associate** button

As shown in the following screenshot, the project is using the **Support Field Configuration Scheme**, which has two configurations:

- The **Bug** and **Sub-task** issue types are using **Default Field Configuration**, since they do not have specific mappings.
- All other issue types, such as Bug and Sub-task, are using their own field configurations as per the scheme's mapping:

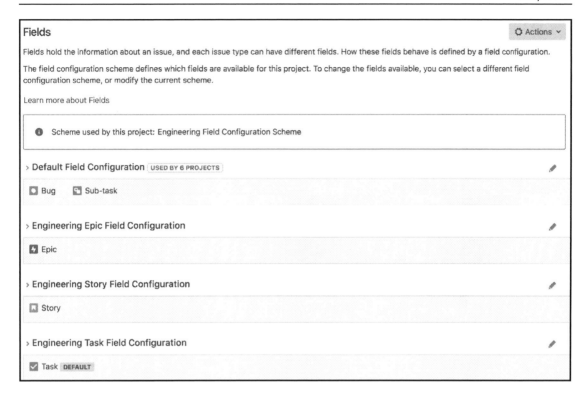

 You can click on each of the field configurations to view their details.

The HR project

Now that you have seen how to manage fields in Jira, it is time to expand our HR project. What we will do this time is add a few new custom fields to help capture some additional useful information. We will also create a customized field configuration specially designed for our HR team. Lastly, we will tie everything together by associating our fields, configurations, and projects through field configuration schemes.

Setting up a custom field

Since you are implementing a project for HR, and we have created two issue types in the last chapter, New Employee and Termination, for the New Employee issue type, we will add a new custom field called Direct Manager, so when everything is completed, the manager can be notified that his/her new team member is ready to start. Since the manager is already in the organization, we will be using a user picker field, so Jira will be able to automatically look up the user for us.

For our Termination issue type, we will also add a new custom field called Last Day, so we know when it will be the last day for the employee. For this field, we will use a date picker, so we can keep the date format consistent.

To create these custom fields, execute the following tasks:

1. Browse to the **View Custom fields** page
2. Click on the **Add Custom Field** button
3. Select the **User Picker** custom field type
4. Name the custom field Direct Manager and click on **Create**
5. Select **HR: Task Management Create Issue Screen** and **HR: Task Management Edit/View Issue Screen**, and click on **Update**
6. Repeat steps 2 to 5, but select the **Date Picker** field type and call it Last Day

Setting up the field configuration

Now that we have our custom fields ready, the next step is to create a new field configuration so that we can specify the behaviors of our custom fields. What we will do here is set both new custom fields as required, so when the issues are entered in Jira, users will have to enter a value for them. But the Direct Manager field should only be required when creating a New Employee issue, and not Termination. To do this, we need to create two field configurations:

1. Browse to the **View Field Configurations** page.
2. Click on the **Add Field Configuration** button.
3. Name the new field configuration New Employee Field Configuration

4. Click on the **Add** button to create a new field configuration. Now that we have our new field configuration, we can start adding configurations to our new custom fields.

5. Click on the **Required** link for the `Direct Manager` custom field.

6. Repeat Steps 2 to 5 to create a new `Termination Field Configuration`, and make the `Last Day` field mandatory.

Setting up a field configuration scheme

We have our custom fields, and have configured the relevant options, created a new field configuration, and set the behavior of our fields. Now it is time to add them to a scheme:

1. Browse to the **View Field Configuration Schemes** page

2. Click on the **Add Field Configuration Scheme** button

3. Name the new field configuration scheme `HR Field Configuration Scheme`, as we will be applying this to our `HR` project

4. Click on the **Add** button to create a new field configuration scheme

With the field configuration scheme created, we can associate the field configurations with their appropriate issue types, `New Employee` and `Termination`:

1. Click on the **Associate an Issue Type with a Field Configuration** button

2. Set the issue type as `New Employee` and the field configuration as `New Employee Field Configuration`

3. Click on the **Add** button to add the association

4. Repeat steps 1 to 3 for the `Termination` issue type and `Termination Field Configuration`

Putting it together

Ok, so we have done all the hard work. We created new custom fields, a new field configuration, and a new field configuration scheme; the final step is to put everything together and see it in action:

1. Browse to the **Project Administration** page for our `HR` project

2. Click on the **Fields** link on the left-hand side and the **Use a different scheme** option from the **Actions** menu

3. Select **HR Field Configuration Scheme** and click on the **Associate** button

Alright, we are all done! You can pat yourself on the back, sit back, and take a look at your hard work in action.

Create a new issue of the `New Employee` type in the `HR` project, and you will see your new custom fields at the bottom of the page. As shown in the following screenshot, the **Direct Manager** field is mandatory and an error message is displayed if we do not select a value for it, while the **Last Day** field is optional:

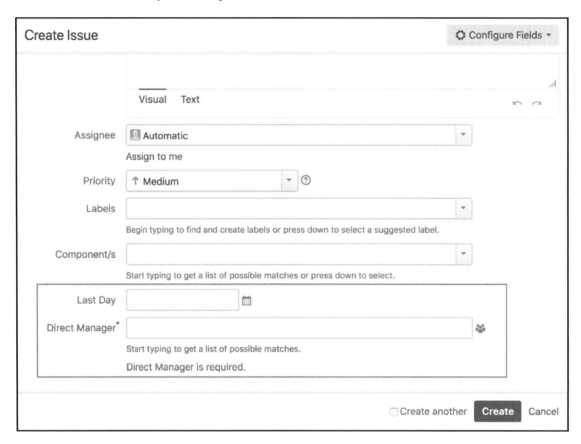

We are seeing both **Direct Manager** and **Last Day** custom fields here because both `New Employee` and `Termination` issue types use the same set of screens. We will look at how to use separate screens in the next chapter. We can, however, also use field configuration to hide the field for the appropriate issue type.

Go ahead and create `New Employee` by filling in the fields. On the **View Issue** page, you will see your new custom fields displayed, along with the values you provide:

Summary

In this chapter, we looked at fields in Jira. We also looked at how Jira is able to extend its ability to capture user data through custom fields. We explored how we can specify different behavior for fields under different contexts through the use of field configurations and schemes.

In the next chapter, we will expand on what we learned about fields by formally introducing you to screens, and will show you how combining fields and screens provides users with the most natural and logical forms to assist them with creating and logging issues.

Screen Management 6

Fields collect data from users, and you have seen how to create your own custom fields from a wide range of field types to address your different requirements. Indeed, data collection is at the center of any information system, but that is only half of the story. How data is captured is just as critical. Data input forms need to be organized so that users do not feel overwhelmed, and the general flow of fields needs to be logically structured and grouped into sections. This is where screens come in.

In this chapter, we will pick up from where we left off in the last chapter and explore the relationship between fields and screens. We will further discuss how you can use screens to customize your Jira to provide users with a better user experience. By the end of the chapter, you will have learned the following:

- What screens are and how to create them
- How to add fields onto screens
- How to break down your screen into logical sections with tabs
- The relationship between screens and issue operations
- How to link screens with projects and issue types
- How to configure project-specific screens as a project administrator

Jira and screens

Before you can start working with screens, you need to first understand what they are and how they are used in Jira. Compared to a normal paper-based form, fields in Jira are like checkboxes and spaces that you have to fill in, and screens are like form documents themselves. When fields are created in Jira, they need to be added to screens in order to be presented to users. Therefore, you can say that screens are like groupings or containers for fields.

In most cases, screens need to be associated with issue operations through what are known as **screen schemes**. Screen schemes map screens to operations, such as creating, viewing, and editing issues, so that you can have different screens for different operations. Screen schemes are then associated with issue type screen schemes, which when applied to projects will map screen schemes to issue types. This lets each issue type in a project have its own set of screens. The only time when a screen will be used directly is when it is associated with a workflow transition. In Jira, a workflow defines the various statuses an issue can go through; for example, an issue can go from open to closed. Transitions are the actions that take the issue from one status to the next, and Jira lets you display a screen as part of the action if you choose to. We will cover workflows in `Chapter 7`, *Workflows and Business Process*.

To help you visualize how screens are used in Jira, Atlassian has provided the following diagram, which summarizes the relationship between fields, screens, and their respective schemes:

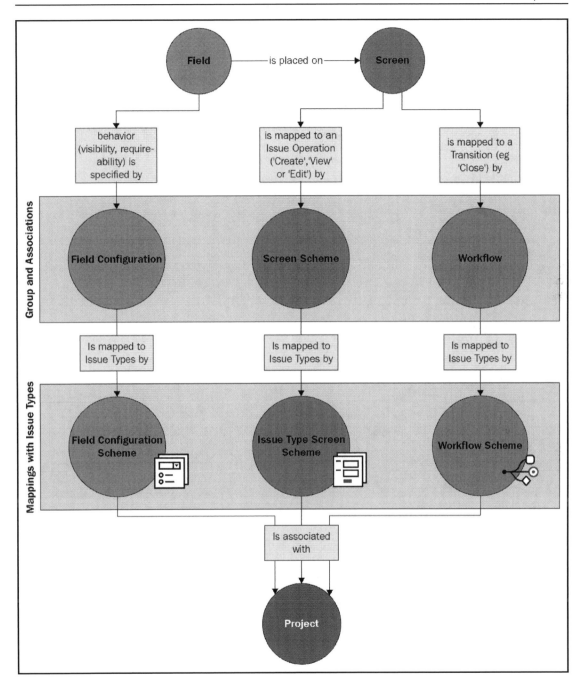

Working with screens

While many other software systems provide users with limited control over the presentation of screens, Jira is very flexible when it comes to screen customizations. You can create your own screens and decide what fields are to be placed on them and their order. You can also decide which screens are to be displayed for major issue operations. In Jira, you can create and design customized screens for the following operations:

- Creating an issue in the create issue dialog box
- Editing an issue when an issue is being updated
- Viewing an issue after an issue is created and is being viewed by users
- Transitioning an issue through a workflow (workflows will be covered in Chapter 7, *Workflows and Business Process*)

Screens are maintained centrally from the administration console, which means you need to be a Jira administrator to create and configure screens. Perform the following steps to access the **View Screens** page:

1. Log in as a Jira administrator user
2. Browse to the Jira administration console
3. Select the **Issues** tab and then the **Screens** option; this will bring up the **View Screens** page

The **View Screens** page lists all the screens that are currently available in your Jira instance. You can select a screen and configure what fields will be on this screen and decide how you can divide a screen into various tabs.

For each of the screens listed here, Jira will also tell you what screen scheme each of the screens is a part of and the workflows that are being used. You have probably noticed that, for screens that are either part of a screen scheme or workflow, there is no **Delete** option available, as you cannot delete screens that are in use. You need to disassociate the screen from screen schemes and/or workflows to delete them, as shown in the following screenshot:

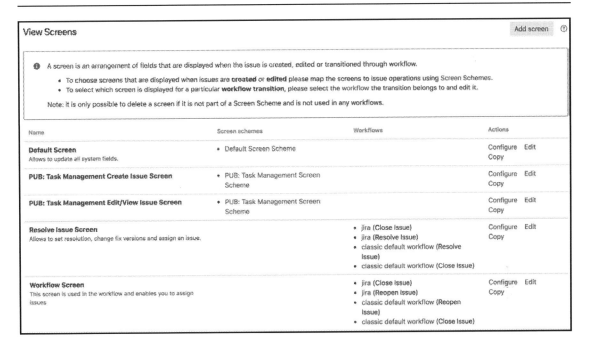

As shown in the preceding screenshot, for each screen you can perform the following operations:

- **Configure**: This configures what fields are to be placed on the screen. It is not to be confused with the **Edit** operation.
- **Edit**: This updates the screen's name and description.
- **Copy**: This makes a copy of the selected screen, including its tabs and field configurations.
- **Delete**: This deletes a screen from Jira. This is only available if it is not being used by a screen scheme or a workflow.

 Screens listed here do not affect Jira Service Desk. We will cover screen and field configuration for Jira Service Desk in Chapter 11, *Jira Service Desk*.

Adding a new screen

Jira comes with three screens by default (as listed in the following bullet list), and every time you create a new project, a new set of screens is created for the project, based on the template you select. These project-specific screens will all have their names starting with the project key, for example, `HD: Task Management View Issue Screen`, where `HD` is the project's key:

- **Default screen**: This screen is used for creating, editing, and viewing issues
- **Resolve Issue screen**: This screen is used when resolving and closing issues
- **Workflow screen**: This screen is used when transitioning issues through workflows (if they are configured to have a screen, such as **Reopen Issue**)

While the default screens and screens automatically created for your projects can cover the most basic requirements, you will soon find yourself outgrowing them, and adjustments will need to be made. For example, if you want to keep certain fields read-only, such as priority, so that they cannot be changed after issue creation, you can achieve this by setting up different screens for creating and editing issues. Another example is having different create and edit screens for different issue types, such as bug and task. In these cases, you will need to create your own screen in Jira using the following steps:

1. Browse to the **View Screens** page.
2. Click on the **Add Screen** button. This will bring up the **Add Screen** dialog box.
3. Enter a meaningful name and description for the new screen. It is a good idea to name your screen after its purpose, for example, `HD: Bug Create Screen`, to indicate that it is the screen to create new bug issues for project `HD`.
4. Click on the **Add** button to create the screen.

At this point, your new screen is blank with no fields in it. You will see in later sections how to add fields onto screens and put them to use.

Editing/deleting a screen

You can edit existing screens by updating their details to help keep your configurations up to date and consistent. Perform the following steps to edit a screen:

1. Browse to the **View Screens** page.
2. Click on the **Edit** link for the screen on which you wish to update. This will take you to the **Edit Screen** page.

3. Update the name and description of the screen.
4. Click on the **Update** button to apply your changes.

To delete an existing screen, it must not be in use by any screen schemes or workflows. If it is associated with a screen scheme or workflow, you will not be able to delete it. You will need to undo the association first. Perform the following steps to delete a screen:

1. Browse to the **View Screens** page.
2. Click on the **Delete** link for the screen you wish to remove. This will take you to the **Delete Screen** page for confirmation.
3. Click on the **Delete** button to remove the screen.

By deleting a screen, you do not delete the fields that are on the screen from the system.

Copying a screen

Screens sometimes can be complex, so creating a new screen from scratch may not be the most efficient method if there is already a similar one available. Just like with many other entities in Jira, you can make a copy of an existing screen, thus cutting down the time that it would otherwise take you to re-add all the fields:

1. Browse to the **View Screens** page.
2. Click on the **Copy** link for the screen you wish to copy. This will take you to the **Copy Screen** page.
3. Enter a new name and description for the screen.
4. Click on the **Copy** button to copy the screen.

Configuring screens

Creating a new screen is like getting a blank piece of paper; the fun part is adding and arranging the fields on the screen. Fields in Jira are arranged and displayed from top to bottom in a single column. You have full control of what fields can be added and in what order they can be arranged.

The only exception to this is for the **View** screen. When you are viewing an issue, fields are grouped together by type. For example, user fields such as reporter and assignee are displayed together on the top right-hand side of the page. Also note that for built-in fields such as **Summary** and **Issue type**, even if you take them off the screen, they will still be displayed when viewing an issue. For these fields, you cannot change their position on the screen.

Jira also allows you to break your screens into tabs or pages within a form, and you can do all of this within a single configuration page. It is this level of flexibility combined with simplicity that makes Jira a very powerful tool.

Perform the following steps to configure an existing screen:

1. Browse to the **View Screens** page
2. Click on the **Configure** link for the screen you wish to configure

On this page, you can do the following:

- Add/remove fields onto the screen
- Rearrange the order of the fields
- Create/delete tabs on the screen
- Move fields from one tab to another

Adding a field to a screen

When you first create a screen, it is of little use. In order for screens to have items to present to the users, you must first add fields onto the screens:

1. Browse to the **Configure Screen** page for the screen you wish to configure.
2. Select the fields you would like to add by typing in the field's name in the **Field name** drop-down list. Jira will auto match the field as you type, as shown in the following screenshot. If you do not see the field you are looking for, make sure it is not already on the screen or on a different tab. Tabs are covered in a later section:

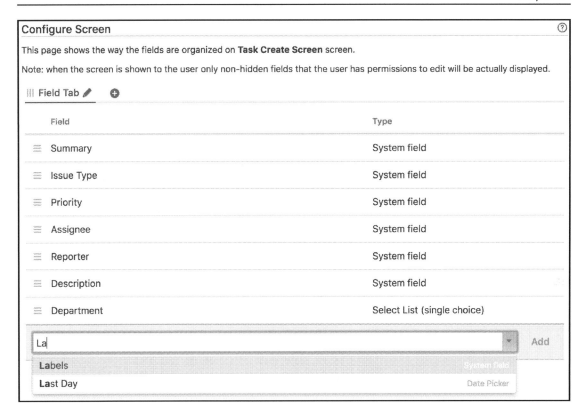

Fields are added to the bottom of the list. You can reorder the list of fields by simply dragging them up and down.

Deleting a field from a screen

Fields can be removed from a screen completely. When a field is taken off, the field will not appear when the screen is presented to the user. There is a subtle difference between deleting a field from a screen and hiding it (as discussed in `Chapter 5`, *Field Management*). Although both actions will prevent the field from showing up, by removing the field issues will not receive a value for that field when they are created. This becomes important when a field is configured to have a default value. When the field is removed from the screen, the issue will not have the default value for the field; while if the field is simply hidden, the default value will be applied.

You will also need to pay close attention when deleting fields from a screen, as there is no confirmation dialog. Make sure that you do not delete required fields, such as **Summary**, from a screen used to create new issues. As seen in `Chapter 5`, *Field Management*, Jira will prevent you from hiding fields that are marked as required, but Jira does not prevent you from taking the required fields off the screen. Therefore, it is possible for you to end up in a situation where Jira requires a value for a field that does not exist on the screen. This can lead to very confusing error messages for end users:

1. Browse to the **Configure Screen** page for the screen you wish to configure
2. Hover your mouse over the field you want to delete and click on the **Remove** button

 When you delete a field from a screen, existing issues will not lose their values for the field. Once you add the field back, the values will be displayed again.

Using screen tabs

For most cases, you will be sequentially adding fields to a screen and users will fill them from top to bottom. However, there will be cases where your screen becomes over complicated and cluttered due to the sheer number of fields you need, or you may simply want to have a way to logically group several fields together and separate them from the rest. This is where tabs come in.

If you think of screens as the entire form a user must fill in, then tabs will be the individual pages or sections that make up the whole document. Tabs go from left to right, so it is a good practice to design your tabs to flow logically from left to right. For example, the first tab can gather general information, such as the summary and description. Subsequent tabs will gather more domain-specific information:

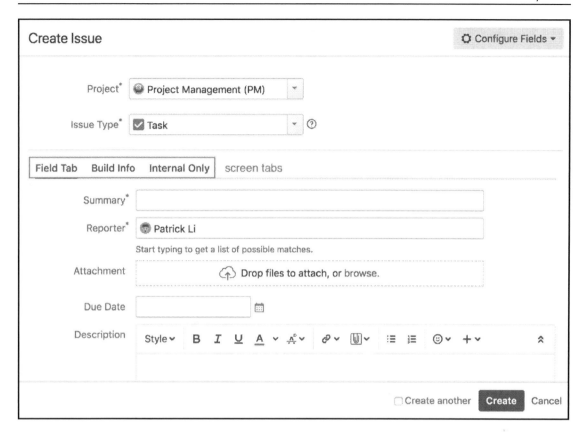

Adding a tab to a screen

You can add tabs to any screen in Jira. In fact, by default, all screens have a default tab called the **Field Tab**, which is used to host all the fields. You can add new tabs to a screen to break down and better manage your screen presentation, as follows:

1. Browse to the **View Screens** page.
2. Click on the **Configure** link for the screen on which you wish to add a new tab.
3. Click on the **Add Tab** link and enter a name for the tab.
4. Click on the **Add** button to create the tab.

Tabs are organized horizontally from left to right. When you add a new tab to the screen, they are appended to the end of the list. You can change the order of tabs by dragging them left and right in the list, as shown in the following screenshot:

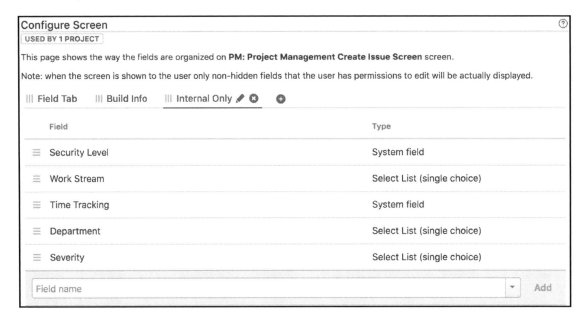

You can also move a field from one tab to another by dragging the field and hovering it over the target tab. This will save you time having to manually remove a field from a tab and then add it to the new tab.

Editing/deleting a tab

Just like screens, you can maintain existing tabs by editing their names and/or removing them from the screen. Perform the following steps to edit a tab's name:

1. Browse to the **View Screens** page
2. Click on the **Configure** link for the screen that has the tab you wish to edit
3. Select the tab by clicking on it
4. Click on the **Edit** icon and enter a new name for the tab
5. Click on the **Save** button to apply the change

When you delete a tab, the fields that are on the tab will be taken off the screen. You will need to re-add or move them to a different tab if you still want those fields to appear on the screen. You cannot delete the last tab on the screen. To delete a tab, perform the following steps:

1. Browse to the **View Screens** page.
2. Click on the **Configure** link for the screen that has the tab you wish to edit.
3. Select the tab by clicking on it.
4. Click on the **Delete** icon. Jira will ask you to confirm whether you want to delete the tab and list all the fields present.
5. Click on the **Delete** button to remove the tab from the screen.

Working with screen schemes

You have seen how we can create and manage screens and how to configure what fields to add to the screens. The next piece of the puzzle is letting Jira know how to choose the screen that has to be displayed for each issue operation.

Screens are displayed during issue operations, and a screen scheme defines the mapping between screens and the operations. With a screen scheme, you can control which issue operations the screen displays, as follows:

- **Create Issue**: This screen is shown when you create a new issue
- **Edit Issue**: This screen is shown when you edit an existing issue
- **View Issue**: This screen is shown when you view an issue

Just like screens, whenever you create a new project in Jira, a new screen scheme is created specifically for your project, and screens are automatically assigned to these issue operations.

The defaults created are usually good enough to get started with; however, there will be times when you don't want certain fields to be available for editing once the issue is created, such as **Issue Type**. You may want to have finer control over the type of issues raised for reporting and statistical measurement reasons, so it is not a good idea to let users freely change the issue type. Another example would be when certain fields are not required during creation because the required information may not be available at the time. Therefore, instead of confusing and/or overwhelming your users, leave those fields out during issue creation and only ask for them to be filled in at a later time when the information becomes available.

As you can see, by dividing the screen into multiple issue operations rather than having the one-screen-fits-all approach, Jira provides you with a new level of flexibility to control and design your screens. As always, if there are no significant differences between the screens, for example, create and edit, it is recommended that you create a base screen and use the **Copy Screen** feature to reduce your workload.

Just like screens, you need to be a Jira administrator to manage screen schemes. Perform the following steps to manage screen schemes:

1. Browse to the Jira administration console.
2. Select the **Issues** tab and then the **Screen schemes** option to bring up the **View Screen Schemes** page:

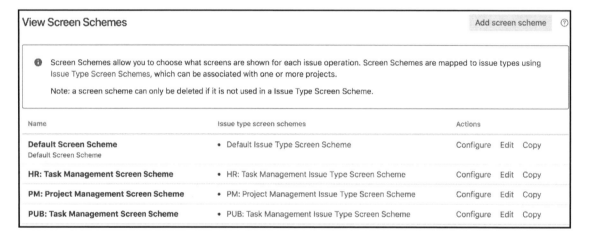

From the **View Screen Schemes** page, you will be able to see a list of all the existing screen schemes, create and manage their configurations, and view their associations with issue type screen schemes (this will be explained in a later section).

Adding a screen scheme

Usually, you will be using the screen scheme created by Jira for your project. However, there will be cases where you need more than one. For example, if you need to display a different set of screens based on the various issue types you have in your project, you will need to create a new screen scheme for each issue type. Perform the following steps to create a new screen scheme:

1. Browse to the **View Screen Schemes** page.
2. Click on the **Add screen scheme** button.
3. Enter a meaningful name and description for the new screen scheme.
4. Select a default screen from the list of screens. This screen will be displayed when no specific issue operation is mapped.
5. Click on the **Add** button to create the screen scheme.

At this stage, the new screen scheme is not in use. This means that it is not associated with any issue type screen schemes yet (issue type screen schemes are covered in later sections).

After a screen scheme is created, it will apply the selected default screen to all the issue operations. We will look at how to associate screens to issue operations in later sections.

Editing/deleting a screen scheme

You can update the details of existing screen schemes, such as their name and description. In order for you to make changes to the default screen selection, you need to configure the screen scheme, which will be covered in later sections. Perform the following steps to edit an existing screen scheme:

1. Browse to the **View Screen Schemes** page.
2. Click on the **Edit** link for the screen scheme you wish to edit. This will take you to the **Edit Screen Scheme** page.
3. Update the name and description with new values.
4. Click on the **Update** button to apply the changes.

Inactive screen schemes can also be deleted. If the screen scheme is active (that is, associated with an issue type screen scheme), then the delete option will not be present. Perform the following steps to delete a screen scheme:

1. Browse to the **View Screen Schemes** page.
2. Click on the **Delete** link for the screen scheme you wish to delete. This will take you to the **Delete Screen Scheme** page.
3. Click on the **Delete** button to confirm that you wish to delete the screen scheme.

Copying a screen scheme

While screen schemes are not as complicated as screens, there will still be times when you would like to copy an existing screen scheme rather than creating one from scratch. You might wish to copy the scheme's screens/issue operations associations, which we will cover in the following section, or make a quick backup copy before making any changes to the scheme.

Perform the following steps to copy an existing screen scheme:

1. Browse to the **View Screen Schemes** page.
2. Click on the **Copy** link for the screen scheme you wish to copy. This will take you to the **Copy Screen Scheme** page.
3. Enter a new name and description for the screen scheme.
4. Click on the **Copy** button to copy the selected screen scheme.

Just like creating a new screen scheme, copied screen schemes are inactive by default.

Configuring a screen scheme

As mentioned earlier, when you create a new screen scheme, it will use the same screen selected as your default screen for all issue operations. Now, if you want to use the same screen to create, edit, and view, then you are all set; there is no need to perform any further configuration to your screen scheme. However, if you need to have different screens displayed for different issue operations, then you will need to establish this association.

When an issue operation does not have an association with a screen, the selected default screen will be applied. If the issue operation is later given in a screen association, then the specific association will take precedence over the general fallback default screen.

The associations between screens and issue operations are managed on a per-screen scheme level. Perform the following steps to configure a screen scheme:

1. Browse to the **View Screen Schemes** page.
2. Click on the **Configure** link for the screen scheme you wish to configure. This will take you to the **Configure Screen Scheme** page.

Associating screens to issue operations

Each issue operation can be associated with one or more issue operations. Perform the following steps to associate an issue operation with a screen:

1. Browse to the **Configure Screen Scheme** page for the screen scheme to be configured.
2. Click on the **Associate an issue operation with a screen** button (as shown in the following screenshot).
3. Select an issue operation to be assigned to a screen.
4. Select the screen to be associated to the issue operation.
5. Click on the **Add** button to create the association.

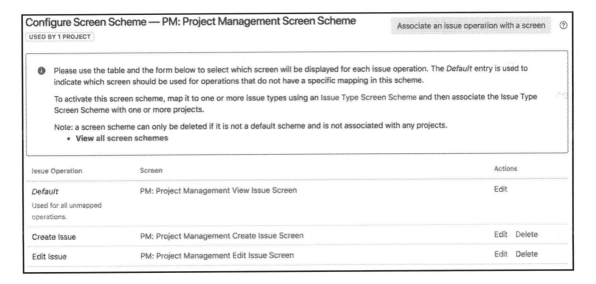

As shown in the preceding screenshot, the **Create Issue** and **Edit Issue** operations are associated with **PM: Project Management Create Issue Screen** and **PM: Project Management Edit Issue Screen**, respectively. Since we do not have a screen associated with the **View Issue** operation, the default association, **PM: Project Management View Issue Screen**, will be used.

Editing/deleting an association

After you create an association for an issue operation, Jira prevents you from creating another association for the same issue operation by removing it from the list of available options. In order to change the association to a different screen, you need to edit the existing association, as follows:

1. Browse to the **Configure Screen Scheme** page for the screen scheme to be configured.
2. Click on the **Edit** link for the association you wish to edit. This will take you to the **Edit Screen Scheme Item** page.
3. Select a new screen to associate with the issue operation.
4. Click on the **Update** button to apply the change.

If you decide that one or more existing associations are no longer needed, then you can delete them from the screen scheme by performing the following steps:

1. Browse to the **Configure Screen Scheme** page for the screen scheme to be configured
2. Click on the **Delete** link for the association you wish to delete

Please note that unlike other similar operations, deleting an issue operation association does not prompt you with a confirmation page. As soon as you click on the **Delete** link, your association will be deleted immediately.

Issue type screen scheme

Screen schemes group screens together and create associations with issue operations. The next piece of the puzzle is to tell Jira to use our screen schemes when creating, viewing, and editing specific types of issues.

We do not directly associate screen schemes to Jira. The reason for this is that Jira has the flexibility to allow you to define this on a per-issue type level. What this means is that, instead of forcing all issue types in a given project to use the same screen scheme, you can actually use different screen schemes for different issue types. This extremely flexible and powerful feature is provided through the **Issue Type Screen Scheme**.

Just like screens and screen schemes, you need to be a Jira administrator to create and manage issue type screen schemes. Perform the following steps to manage issue type screen schemes:

1. Browse to the Jira administration console.
2. Select the **Issues** tab and then the **Issue Type Screen Schemes** option to bring up the **Issue Type Screen Schemes** page:

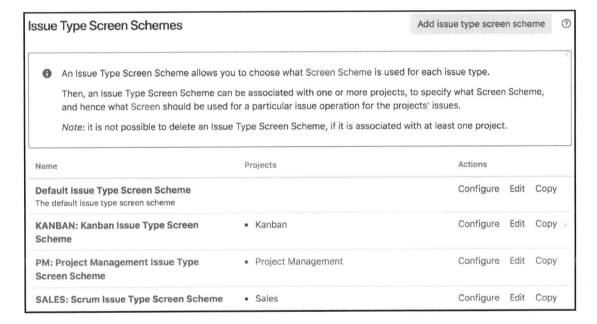

Adding an issue type screen scheme

Just as with a screen scheme, Jira will automatically create an issue type screen scheme when you create your project. Since one project can only have one issue type screen scheme associated with it, usually you will not need to create new ones yourself. However, there might be a time when you want to create a new scheme, such as experimenting with some new configurations while still wanting to keep the existing one untouched in case of a roll-back.

Perform the following steps to create a new issue type screen scheme:

1. Browse to the **Issue Type Screen Schemes** page
2. Click on the **Add issue type screen scheme** button
3. Enter a name and description for the new issue type screen scheme
4. Select a default screen scheme from the list of screen schemes
5. Click on the **Add** button to create the issue type screen scheme

That's right, you guessed it! The new issue type screen scheme is not yet in use. It will only become active once it is applied to one or more projects, which we will look at shortly.

Editing/deleting an issue type screen scheme

You can make updates to an existing issue type screen scheme's name and description. To change its screen scheme/issue type association details, you need to configure the issue type screen scheme, which will be covered in later sections. Perform the following steps to update an issue type screen scheme:

1. Browse to the **Issue Type Screen Schemes** page.
2. Click on the **Edit** link for the issue type screen scheme you wish to edit. This will take you to the **Edit Issue Type Screen Scheme** page.
3. Update the name and description with new values.
4. Click on the **Update** button to apply the changes.

Just as with all of the other schemes in Jira, you cannot delete issue type screen schemes that are in use. You will have to make sure that no project uses it before Jira allows you to delete the scheme. To delete issue type screen schemes, perform the following steps:

1. Browse to the **Issue Type Screen Schemes** page.
2. Click on the **Delete** link for the issue type screen scheme you wish to delete. This will take you to the **Delete Issue Type Screen Scheme** page.
3. Click on the **Delete** button to remove the issue type screen scheme.

Copying an issue type screen scheme

Issue type screen scheme cloning is also available in Jira. You can easily make copies of existing issue type screen schemes. One very useful application of this feature is that it enables you to make backup copies before experimenting with new configurations. Note that copying the issue type screen scheme does not back up the screen schemes and screens that it contains.

Perform the following steps to copy an existing issue type screen scheme:

1. Browse to the **Issue Type Screen Schemes** page.
2. Click on the **Copy** link for the issue type screen scheme you wish to copy. This will take you to the **Copy Issue Type Screen Scheme** page.
3. Enter a new name and description for the issue type screen scheme.
4. Click on the **Copy** button to copy the selected scheme.

Newly created issue type screen schemes are inactive by default, while cloned schemes are not used by any projects.

Configuring an issue type screen scheme

By creating new issue type screen schemes, you can establish new associations between screen schemes and issue types. These associations are what tie projects and issue types to individual screens.

Each issue type screen scheme needs to be configured separately, and the associations created are specific to the configured scheme:

1. Browse to the **Issue Type Screen Schemes** page.
2. Click on the **Configure** link for the issue type screen scheme you wish to configure. This will take you to the **Configure Issue Type Screen Scheme** page.

Associating issue types to screen schemes

Jira determines which screen scheme to use for an issue type by establishing an association between screen schemes and issue types. Each issue type can have only one screen scheme associated with it. However, each screen scheme can be associated with more than one issue type.

Perform the following steps to add a new association:

1. Browse to the **Configure Issue Type Screen Scheme** page for the issue type screen scheme you wish to configure.
2. Click on the **Associate an Issue Type with a Screen Scheme** button.
3. Select the issue type to add an association.
4. Select the screen scheme to be associated with the issue type.
5. Click on the **Add** button to create the association:

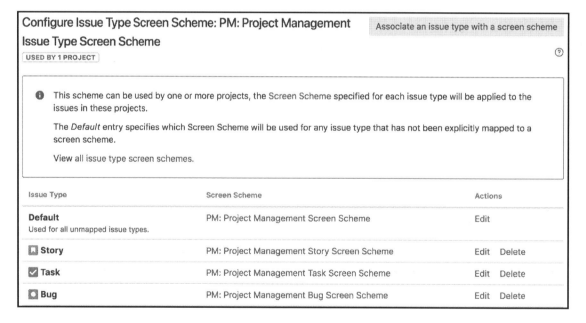

As shown in the preceding screenshot, the **Story**, **Task**, and **Bug** issue types are explicitly associated with **PM: Project Management Story Screen Scheme**, **PM: Project Management Task Screen Scheme**, and **PM: Project Management Bug Screen Scheme**, respectively. All other issues types, such as **Sub-task**, will be associated with the default **PM: Project Management Screen Scheme**.

Editing/deleting an association

You can update existing associations, such as the **Default** association, which is created automatically when you create a new issue type screen scheme:

1. Browse to the **Configure Issue Type Screen Scheme** page for the issue type screen scheme to be configured
2. Click on the **Edit** link for the association you wish to edit. This will take you to the **Edit Issue Type Screen Scheme** page
3. Select a new screen scheme to associate it with the issue type
4. Click on the **Update** button to apply the change

You can also delete existing associations for issue types if you no longer need them to be explicitly set. However, you cannot delete the **Default** association, since it is used as a catch for all of the issue types that do not have an association defined. This is important because, while you may have created associations for all of the issue types right now, you might add new issue types down the line and forget to create associations for them. To delete an association, perform the following steps:

1. Browse to the **Configure Issue Type Screen Scheme** page for the issue type screen scheme to be configured
2. Click on the **Delete** link for the association you wish to delete

Just like associations in screen schemes, you will not be taken to a confirmation dialog, and the association will be deleted immediately.

Associating an issue type screen scheme with a project

Perform the following steps in order to activate your new issue type screen scheme, which will display your new screens for the different issue operations:

1. Browse to the target project's administration page.
2. Click on the **Screens** option from the left panel.

3. Select the **Use a different scheme** option from the **Actions** menu:

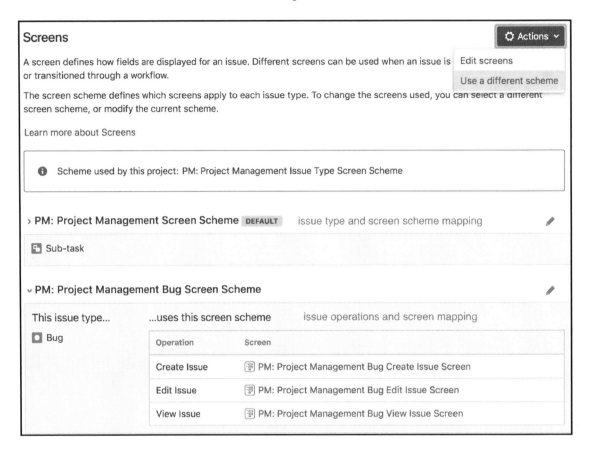

4. Select the issue type screen scheme from the **Scheme** select list.
5. Click on the **Associate** button.

Once you associate the issue type screen scheme with the project, Jira will show you the details of the mapping, as shown in the preceding screenshot.

Delegating screen management

Managing screen configurations used to be centrally controlled by the Jira administrator. The project administrator can only select what issue type screen scheme to use, but if modifications need to be made for the screens, the Jira administrator will need to be involved. This often creates a bottleneck for simple things, such as adding or removing a field from a screen, especially for large organizations that have many projects but only a few Jira administrators.

Jira now has a new feature called **Extended Project Administration**, which empowers project administrators by allowing them to make changes to screens used by their projects.

 Extended project administration is controlled via permission settings, which we will cover in `Chapter 9`, *Securing Jira*.

There are, however, some restrictions for this, as listed here:

- The screen must not be a default system screen
- The screen must already be used by the project
- The screen must not be shared with any other projects, or used as a transition screen in workflows
- Only fields that already exist in the system can be added to a screen

Essentially, this means that you, as a project administrator, can only make changes to screens that are dedicated a single project. If the screen is shared with multiple projects, you will still need help from a Jira administrator. To make changes to screens for your project as a project administrator, perform the following steps:

1. Browse to the target project's administration page
2. Click on the **Screens** option from the left panel
3. Expand the issue type and select the screen you want to configure

If the screen can be configured, you should see something similar to the following screenshot, where you have the familiar screen configuration page, but now shown inside the context of a project:

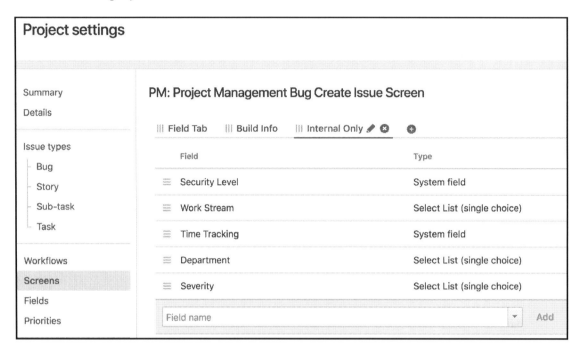

The HR project

Armed with the new knowledge that you have gathered in this chapter, together with fields from the previous chapter, it is time for you to further customize your Jira to provide a better user experience through the presentation.

What we will do this time is create new screens and apply them to our HR project. We want to separate f the new fields are showing
generic fields from our specialized custom fields designed for handling employee onboarding and termination. We also want to apply the changes to the issues of the New Employee and Termination types only, and not affect the other issue types in the project. As with any changes to be done on a production system, it is critical that you have a backup of your current data before applying changes.

Setting up screens

In Chapter 5, *Field Management*, you created a few custom fields specifically designed for our HR team. The problem we had is that all of the new fields show up for both the New Employee and Termination issue types, regardless of whether they are applicable, and this is because both issue types use the same set of screens.

To address this, we will create two new sets of screens, one for New Employee and one for Termination. The default one can be left for other issue types we have in the project, such as Task.

The easiest way to do this will be to clone the existing screens, so we do not have to manually add all of the fields, and avoid forgetting to add a field by accident. To create screens for each issue type, perform the following steps:

1. Browse to the **View Screens page** and click on the **Copy** link for HR: Task Management Create Issue Screen
2. Name the new screen HR: Create/View New Employee Screen
3. Click on the **Copy** button to create the screen

Now that we have our new screen, it is time to configure its fields using the following steps. Since this screen is for creating New Employee issues, we do not need the Last Day field:

1. Click on the **Configure** link of our new HR: Create/View New Employee Screen.
2. Remove the Last Day field by hovering over it and clicking its **Remove** button.

Just to spice things up a bit, we can also create a new People tab and move all people-related fields, such as the Assignee, Reporter, and Direct Manager fields, onto that tab.

We created and configured our create screen. Our new edit screen is going to look very similar to this with just a few modifications. We want to remove the **Issue Type** field, since we do not want users to change the issue type after it is created:

1. Click on the **Copy** link for HR: Create/View New Employee Screen
2. Name the new screen HR: Edit New Employee Screen
3. Click on the **Copy** button to create the new screen
4. Remove the **Issue Type** field

Repeat the steps to create a new set of screens for the Termination issue type. This time, instead of removing the Last Day field, remove the Direct Manager field from both screens.

Setting up screen schemes

With the screens created and configured, we now need to link them up with issue operations, so that Jira will know on which action the new screens will be displayed, using the following steps:

1. Browse to the **View Screen Schemes** page and click on **Add Screen Scheme**
2. Name the new screen scheme`HR: New Employee Screen Scheme`
3. Select `HR: Create/View New Employee Screen` as the default screen
4. Click on the **Add** button to create the screen scheme

With our screen scheme in place, it is time to link up our screens with their respective issue operations, as follows:

1. Click on the **Associate an Issue Operation with a Screen** button
2. Select `HR: Edit New Employee Screen` for the **Edit Issue** operation

Since we assigned `HR: Create/View New Employee Screen` to **Default**, this screen will be applied to unmapped operations—**Create Issue** and **View Issue**. There are no differences if you choose to explicitly set the mappings for the preceding two operations.

We have created the screen scheme for the `New Employee` issue type. Now repeat the same steps for the `Termination` issue type.

Setting up issue type screen schemes

Now you need to tell Jira which issue type to apply the screen scheme to the one you just created. Since Jira has already created an issue type screen scheme for our project, we just need to configure it to use our new screen schemes for the appropriate issue types:

1. Browse to the **Issue Type Screen Schemes** page and click on the **Configure** link for **HR: Task Management Issue Type Screen Scheme**
2. Click on the **Associate an Issue Type with a Screen Scheme** button
3. Select **New Employee** for **Issue Type**
4. Select **HR: New Employee Screen Scheme** as the screen scheme to be associated
5. Click on the **Add** button to create the association

This will ensure that issues of the `New Employee` type will have your new screens applied, while issues of other types will not be affected. Now repeat the steps to associate the `Termination` issue type with its screen scheme.

Putting it together

Since we are reusing the existing issue type screen scheme by associating various issue types to our new screen schemes, we do not need to make any additional changes. However, if you created a new issue type screen scheme instead, you will need to associate it with the HR project.

You can now take a look at your hard work and see your custom screens, fields and tabs all working nicely together to present you with a custom form for collecting user data. Let's go ahead and create a new incident and see what your newly customized **Create Issue** screen will look like, as shown in the following screenshot:

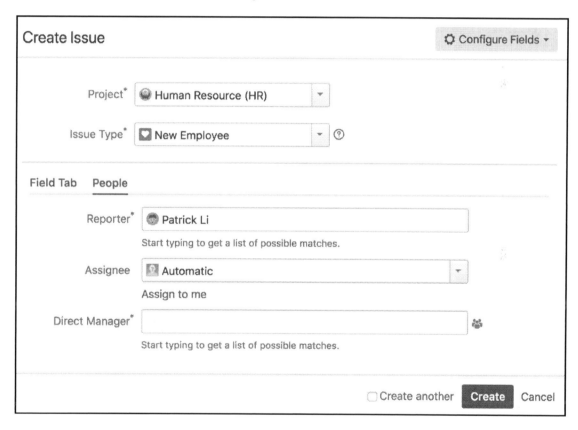

As you can see, the Last Day field is no longer showing on the screen when you create a New Employee issue, and our people-related fields are now showing on the new **People** tab. If you create a new Termination issue, the **Direct Manager** field will not display.

Summary

In this chapter, we looked at how Jira structures its presentation with screens. We looked at how screens are used in Jira via screen schemes, which map screens to issue operations. We also looked at how issue type screen schemes are then used to map screen schemes to issue types. Therefore, for any given project, each issue type can have its own set of screens for create, edit, and view. We also discussed how screens can be broken down into tabs to provide a more logical grouping of fields, especially when your screen starts to have a lot of fields on it.

Together with custom fields, which we saw in the previous chapter, we can now create effective screen designs to streamline our data collection. In the next chapter, we will delve into one of the most powerful features in Jira **workflows**.

7
Workflow and Business Process

In the previous chapters, we learned some of the basics of Jira and how to customize its data collection and presentation with custom fields and screens. In this chapter, we will dive in and take a look at workflows, one of the cores and most powerful features in Jira.

A workflow controls how issues in Jira move from one status to another, as they are being worked on, often passing from one assignee to another. Unlike many other systems, Jira allows you to create your own workflows to resemble your processes.

By the end of this chapter, you will have learned the following:

- What a workflow is and what it consists of
- The relationship between workflows and screens
- What statuses, transitions, conditions, validators, and post functions are
- How to create your own workflow with the workflow designer
- How to associate a workflow with projects

Mapping business processes

It is often said that a good software system is one that adapts to your business and not one that requires your business to adapt to the software. Jira is an excellent example of the former. The power of Jira is that you can easily configure it to model your existing business processes through the use of workflows.

A business process flow can often be represented as a flow chart. For example, a typical document approval flow may include tasks such as document preparation, document review, and document submission, where the user needs to follow these tasks in sequential order. You can easily implement this as a Jira workflow. Each task will be represented as a workflow status with transitions guiding you on how you can move from one status to the next. In fact, when working with workflows, it is often a good approach to first draft out the logical flow of the process as a flow chart and then implement this as a workflow. As we will see, Jira provides many tools to help you visualize your workflows.

Now that we have briefly seen how you can map a normal business process to a Jira workflow, it is time to take a closer look at the components of a workflow, and how you can create your own workflows.

Understanding workflows

A workflow is what Jira uses to model business processes. It is a flow of statuses (steps) that issues go through one by one, with paths between the statuses (transitions). All issues in Jira have a workflow applied, based on their issue type and project. Issues move through workflows from one status (for example, **OPEN**) to another (for example, **CLOSED**). Jira allows you to visualize and design workflows as a diagram, as shown in the following diagram:

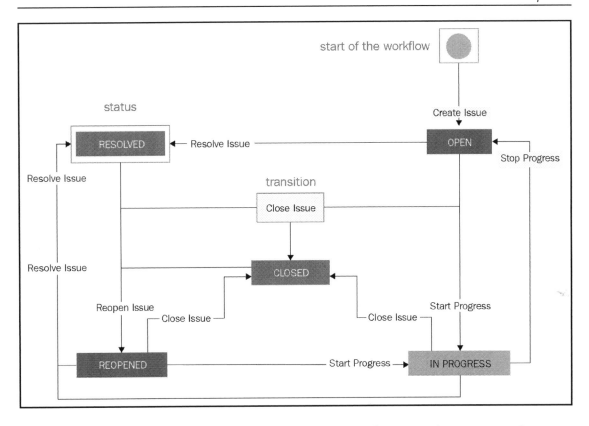

The preceding diagram shows a simple workflow in Jira. The rectangles represent the statuses, and the arrow lines represent transitions that link statuses together. As you can see, this looks a lot like a normal flow chart depicting the flow of a process.

Also, notice that statuses have different colors. The color of a status is determined by the category it belongs to. There are three categories—to do (blue), in progress (yellow), and done (red). Categories help you to easily identify where along the workflow an issue is, by using color as an indicator.

Issues in Jira, starting from when they are created, go through a series of steps identified as **issue statuses**, such as **IN PROGRESS** and **CLOSED**. These movements are often triggered by user interactions. For example, when a user clicks on the **Start Progress** link, the issue is transitioned to the **IN PROGRESS** status, as shown in the following screenshot:

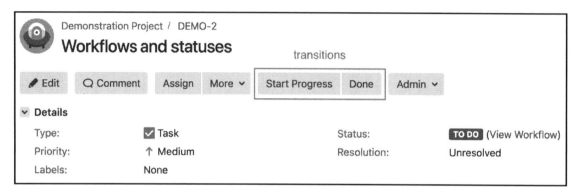

There is a definitive start of a workflow, which is when the issue is first created, but the end of a workflow can sometimes be ambiguous. For example, in the default workflow, issues can go from **OPEN** to **CLOSED** to **REOPENED** and back to **CLOSED**. By convention, when people talk about the end of a workflow, they are usually referring to a status named **CLOSED** or the status where issues are given a **resolution**. Once a resolution is given, the issue comes to a logical end. Several built-in features of Jira follow this convention; for example, issues with resolutions set will not be displayed on the **Assigned to Me** list on the home page.

 When work for an issue is completed, it should be given a resolution.

Managing workflows

Workflows are controlled and managed centrally from the Jira administration console, so you need to be an administrator to create and configure workflows. To manage workflows, perform the following steps:

1. Log in to Jira as a Jira administrator.
2. Browse to the Jira administration console.

3. Select the **Issues** tab and then the **Workflows** option. This will bring up the **View Workflows** page:

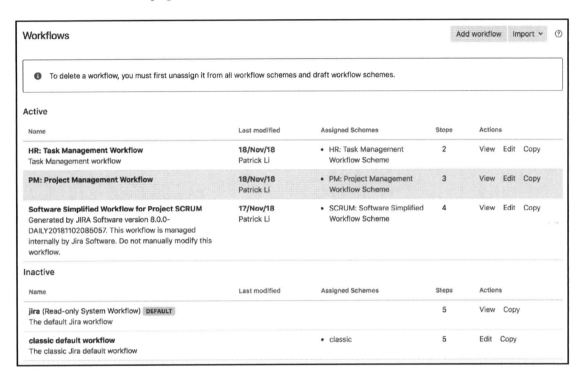

From the **View Workflows** page, you will be able to manage all the available workflows and create new workflows. The page is divided into two sections, **Active** and **Inactive**. Active workflows are being used by projects, and inactive ones are not. By default, the **Inactive** section is collapsed, so the page will only display active workflows. The preceding screenshot shows the **Inactive** section being expanded.

Jira comes with a default read-only workflow called **jira**, mostly used to remain backward-compatible with existing projects, and is applied to projects that do not have any specific workflow applied. For this reason, you cannot edit or delete this workflow. New projects will have their own workflows created based on the template selected. These project-specific workflows will have their names start with the project key, followed by the project's template, such as HR: Task Management Workflow.

Issue statuses

In a Jira workflow, an issue status represents a state in the workflow for an issue. It describes the current status of the issue. If we compare it to a flow chart, the statuses will be the rectangles and, in the diagram, they indicate the current status of the issue along the process. Just as a task can only be in one stage of a business process, an issue can be in only one status at any given time; for example, an issue cannot be both open and closed at the same time.

There is also a term called **step**, which is the workflow terminology for statuses. Since Jira has simplified its workflow administration, step and status can be used interchangeably. For consistency, we will be using the term **status** in this book, unless a separation needs to be made in special cases.

Transitions

Statuses represent stages in a workflow, and the path that takes an issue from one status to the next is known as a **transition**. A transition links two statuses together. A transition cannot exist on its own, meaning it must have a start and finish status and can only have one of each. This means that a transition cannot conditionally split off to different destination statuses. Transitions are also one-way only. This means that if a transition takes an issue from status *A* to status *B*, you must create a new transition if you want to go back from status *B* to status *A*.

Transitions have several components. They are as follows:

- **Conditions**: Criteria must be met before the transition is available (visible) for users to execute. It is usually used to control permissions around how users can execute the transition.
- **Validators**: These are the verifications that must pass before the transition can be executed. They are usually used together with transition screens.
- **Post functions**: These are additional functions to be performed as part of the transition process.
- **Transition screen**: This is an optional screen to be displayed when a user is executing the transition. It is usually used to capture additional information as a part of the transition.

- **Triggers**: If you have integrated Jira with other development tools such as Bitbucket or GitHub, triggers can automatically execute the transition when an event happens, such as the creation of a new branch or when someone makes a code commit.

 A common trick is to create a transition that links back to itself. Since a transition can have its own screen and execute some business logic via post functions, you can use this kind of transition as a trigger in the UI to show a screen or run a post function without having to create complex customizations.

Each of the first three components defines the behavior of the transitions, allowing you to perform pre- and post-validations, as well as post-execution processing on the transition execution. We will discuss these components in depth in the following sections.

Triggers

As described earlier, Jira needs to be integrated with one of the following systems before you can start using triggers:

- Atlassian Bitbucket
- Atlassian FishEye/Crucible
- GitHub

Triggers will listen for changes from the integrated development tools, such as code commits, and when these happen, the trigger will automatically execute the workflow transition. Note that all permissions are ignored when this happens.

Conditions

Sometimes, you might want to have control over who can execute a transition or when a transition can be executed. For example, a transition to authorize an issue should be restricted to users in the manager's group so normal employees will not be able to authorize their own requests. This is where conditions come in.

Conditions are criteria that must be fulfilled before the user is allowed to execute the transition. If the conditions on transitions are not met, the transition will not be available to the user when viewing the issue. The following table shows a list of conditions that are shipped with Jira; additional conditions can be added via third-party add-ons:

Condition	Description
Code Committed Condition	This allows the transition to execute only if the code has/has not (depending on configuration) been committed against this issue.
Hide Transition from User	This will hide the transition from all users, and it can only be triggered by post functions. This is useful in situations where the transition will be triggered as part of an automated process rather than manually by a user.
No Open Reviews Condition	This allows a transition to execute only if there are no related open crucible reviews.
Only Assignee Condition	This only allows the issue's current assignee to execute the transition.
Only Reporter Condition	This only allows the issue's reporter to execute the transition.
Permission Condition	This only allows users with the given permission to execute the transition.
Sub-Task Blocking Condition	This blocks the parent issue transition depending on all its subtasks' statuses.
Unreviewed Code Condition	This allows a transition to execute only if there are no unviewed change sets related to this issue.
User Is In Group	This only allows users in a given group to execute the transition.
User Is In Group Custom Field	This only allows users in a given group custom field to execute a transition.
User Is In Project Role	This only allows users in a given project role to execute a transition.

Validators

Validators are similar to conditions, but they validate certain criteria before allowing the transition to complete. While conditions will hide a workflow transition from the user if its criteria are not met, validators will allow the user to see the transition but not allow the transition to execute if its criteria are not met.

The most common use case for validators is to validate the user input during a transition. For example, you can validate if the user has entered data for all fields presented on the workflow screen. The following table shows a list of validators that come shipped with Jira; additional validators can be added via third-party add-ons:

Validator	Description
Permission Validator	This validates that the user has the selected permission. This is useful when checking whether the person who has executed the transition has the required permissions.
User Permission Validator	This validates that the user has the selected permission where the `OSWorkflow` variable holding the username is configurable. This is obsolete.

Post functions

As the name suggests, post functions are functions that occur after (post) a transition has been executed. This allows you to perform additional processes once you have executed a transition. Jira makes heavy uses of post functions internally to perform a lot of its functions. For example, when you transition an issue, Jira uses post functions to update its search indexes so your search results will reflect the change in issue status.

If a transition has failed to execute (for example, failing validation from validators), post functions attached to the transition will not be triggered. The following table shows a list of post functions that come shipped with Jira, and additional post functions can be added via third-party add-ons:

Post function	Description
Assign to Current User	This assigns the issue to the current user if the current user has the assignable user permission.
Assign to Lead Developer	This assigns the issue to the project/component lead developer.
Assign to Reporter	This assigns the issue to the reporter.
Create Perforce Job Function	This creates a perforce job (if required) after completing the workflow transition.
Notify HipChat	This sends a notification to one or more HipChat rooms.
Trigger a Webhook	If this post function is executed, Jira will post the issue content in JSON format to the URL specified.
Update Issue Field	This updates a system field such as Summary to a given value.

Using the workflow designer

Jira comes with a simple to use, drag-and-drop tool called the **workflow designer**. This helps you create and configure workflows. If you are familiar with diagramming tools such as Microsoft Visio, you will feel right at home. There is also another older option, called the text mode, available. However, since the designer is easier and has more features, we will focus on using the designer in this book.

As your workflow becomes more complicated, the text mode can be a better option to manage statuses and transitions in the workflow.

The workflow designer is shown in the following screenshot. You have the workflow layout in the main panel and a few controls on top, namely the **Add status** and **Add transition** buttons. Note that the **Diagram** option is selected. If you click on the **Text** option, Jira will change to the old authoring tool:

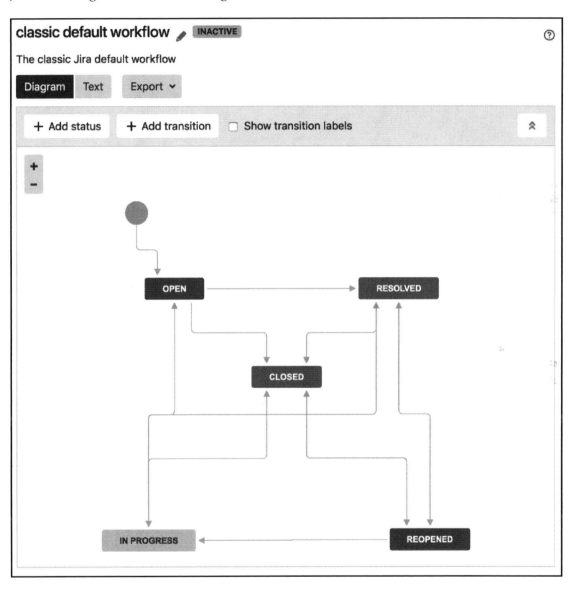

From the workflow designer, you can drag and rearrange the statuses and transitions. Clicking on each will open up its property window, as shown in the following screenshot, where the **Resolved Issue** transition is selected. From here, we can view and update its properties, such as conditions and validators:

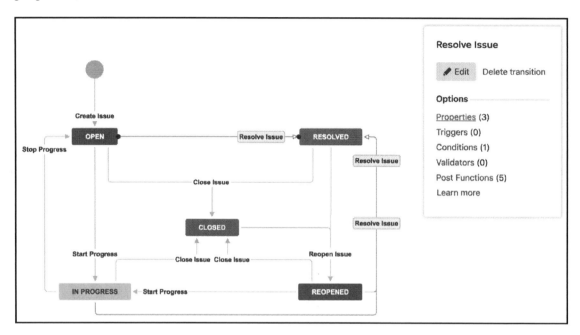

Authoring a workflow

So, let's take a look at how to create and set up a new workflow in Jira. To create a new workflow, all you need is a name and description:

1. Browse to the **View Workflows** page
2. Click on the **Add Workflow** button
3. Enter a name and description for the new workflow in the **Add Workflow** dialog
4. Click on the **Add** button to create the workflow

The newly created workflow will only contain the default create and open statuses, so you will need to configure it by adding new statuses and transitions to make it useful. Let's start with adding new statuses to the workflow using the following steps:

1. Click on the **Add status** button.
2. Select an existing status from the drop-down list. If the status you need does not exist, you can create a new status by entering its name and pressing the *Enter* key on your keyboard.
3. Check the **Allow all statuses to transition to this one** option if you want users to be able to move the issue into this status regardless of its current status. This will create a global transition, which is a convenient option, so you do not have to manually create multiple transitions for the status.

> If the global status is not representing a Done or Closed status, it is often a good idea to add a Clear Resolution post function to make sure the resolution field is always cleared when issue is transitioned into the status.

4. Click on the **Add** button to add the status to your workflow. You can repeat these steps to add as many statuses as you want to your workflow:

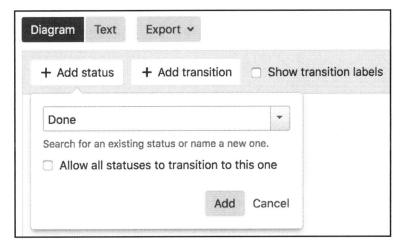

> Try to reuse existing statuses if possible so that you do not end up with many similar statuses to manage.

Now that the statuses are added to the workflow, they need to be linked with transitions so that issues can move from one status to the next. There are two ways to create a transition:

- Click on the **Add transition** button

- Select the originating status and then click and drag the arrow to the destination status

Both options will bring up the **Add Transition** dialog, as shown in the following screenshot:

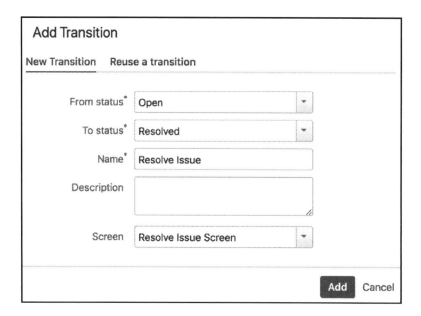

From the preceding screenshot, you can choose to either create a new transition with the **New Transition** tab, or use an existing transition with the **Reuse a transition** tab.

When creating a new transition, you will need to configure the following:

- **From status**: This is the originating status. The transition will be available when the issue is in the selected status.
- **To status**: This is the destination status. Once the transition is executed, the issue will be put into the selected status.
- **Name**: This is the name of the transition. This is the text that will be displayed to users. Since transitions are actions performed by users, it is usually a good idea to name your transitions starting with a verb, such as `Close Issue`.

- **Description**: This is an optional text description showing the purpose of this transition. This will not be displayed to users.
- **Screen**: This is an optional intermediate screen to be displayed when users execute the transition. For example, you display a screen to capture additional data as part of the transition. If you do not select a screen, the transition will be executed immediately. The following screenshot shows a workflow screen:

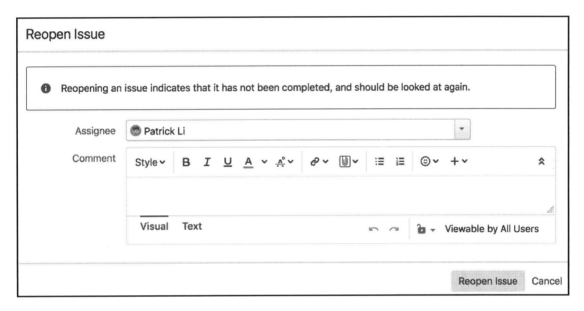

If you want to reuse an existing transition, simply click on the **Reuse a transition** tab, **From status** and **To status**, and **Transition to reuse**, as shown in the following screenshot:

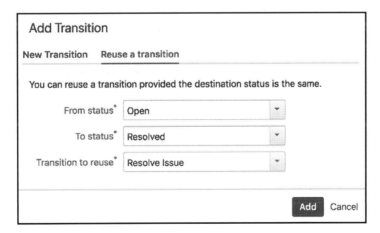

Note that Jira will only list valid transitions based on the **To status** selection.

You might be wondering when you should create a new transition and when you should reuse an existing transition. The big difference between the two is that when you reuse a transition, all instances of the reused transition, also known as the **common transition**, will share the same set of configurations, such as conditions and validators. Also, any changes made to the transition will be applied to all instances. A good use case for this is when you need to have multiple transitions with the same name and setup, such as `Close Issue`; instead of creating separate transitions each time, you can create one transition and reuse it whenever you need a transition to close an issue. Later on, if you need to add a new validator to the transition to validate additional user input, you will only need to make the change once, rather than multiple times for each `Close Issue` transition.

Another good practice to keep in mind is to not have a *dead end* state in your workflow, for example, by allowing closed issues to be reopened. This will prevent users from accidentally closing an issue and not being able to correct the mistake.

One thing people often overlook is that you can change the status an issue is transitioned to when it is first created. By default, an issue is placed in the open status as soon as it is created. While this makes sense for most cases, you can actually change that. For example, you might want all your issues to be in a waiting status and transition to open only after someone has reviewed it. You can also make changes to the default **Create Issue** transition. By doing so, you can influence the issue creation process. For example, you can add a validator to it to add additional checking before an issue is allowed to be created, or add a post function to perform additional tasks as soon as an issue is created.

Now that we have seen how to add new statuses and transitions to a workflow, let's look at adding triggers, conditions, validators, and post functions to a transition.

Adding a trigger to transitions

You can only add triggers to transitions if Jira is integrated with at least one of the supported development tools. With triggers, you can automate some of your DevOps flow, such as automatically transitioning an issue into an **In Review** status when a pull request is created. To add triggers, perform the following steps:

1. Select the transition you want to add triggers to.
2. Click on the **Triggers** link.

3. Click on the **Add trigger** button. If you do not have any integrated development tools, this button will be disabled:

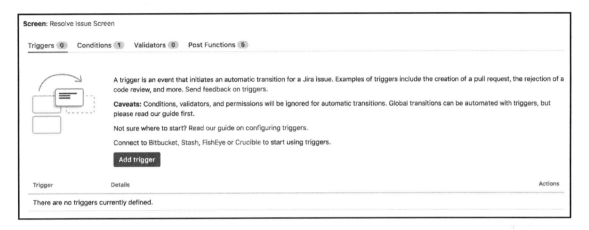

4. Select the trigger you want to add and click on the **Next** button.
5. Confirm the trigger source is detected and click on the **Add trigger** button.

Adding a condition to transitions

New transitions do not have any conditions by default. This means that anyone who has access to the issue will be able to execute the transition. Jira allows you to add any number of conditions to the transition:

1. Select the transition you want to add conditions to.
2. Click on the **Conditions** link.
3. Click on the **Add condition** link. This will bring you to the **Add Condition To Transition** page, which lists all the available conditions you can add:

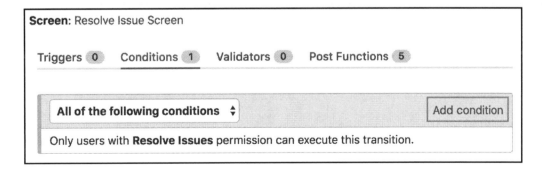

4. Select the condition you want to add.

5. Click on the **Add** button to add the condition.

6. Depending on the condition, you may be presented with the **Add Parameters To Condition** page where you can specify the configuration options for the condition. For example, the **User Is In Group** condition will ask you to select the group to check against, shown as follows:

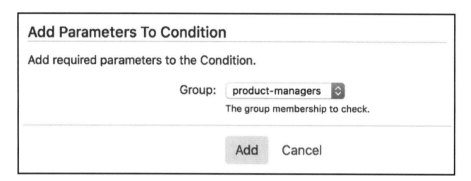

Newly added conditions are appended to the end of the existing list of conditions, creating a **condition group**. By default, when there is more than one condition, a logical AND is used to group the conditions. This means that all conditions must pass for the entire condition group to pass. If one condition fails, the entire group fails, and the user will not be able to execute the transition. You can switch to use the logical OR condition, which means only one of the conditions in the group needs to pass for the entire group to pass. This is a very useful feature as it allows you to combine multiple conditions to form a more complex logical unit.

For example, the **User Is In Group** condition lets you specify a single group, but with the AND operator, you can add multiple **User Is In Group** conditions to ensure that the user must exist in all the specific groups to be able to execute the transition. If you use the OR operator, then the user will only need to belong to one of the listed groups. The only restriction to this is that you cannot use both operators for the same condition group.

 One transition can only have one condition group, and each conditional group can only have one logical operator.

Adding a validator to transitions

Like conditions, transitions, by default, do not have any validators associated. This means that transitions are completed as soon as they are executed. You can add validators to transitions to make sure that executions are only allowed to be complete when certain criteria are met. Use the following steps to add a validator to a transition:

1. Select the transition you want to add conditions to.
2. Click on the **Validators** link.
3. Click on the **Add validator** link. This will bring you to the **Add Validator To Transition** page, which lists all the available validators you can add:

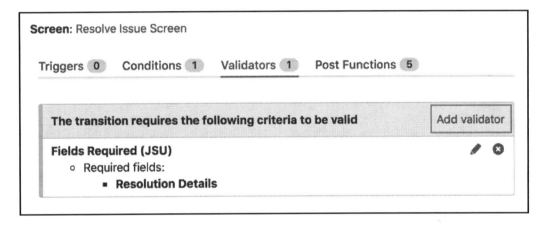

4. Select the validator you want to add.
5. Click on the **Add** button to add the validator.

6. Depending on the validator, you may be presented with the **Add Parameters To Validator** page where you can specify configuration options for the validator. The following screenshot shows an example from the **Fields required** validator:

Add Parameters To Validator

Add required parameters to the Validator.

Fields required:

Available fields:	Required fields:
Last Day Last Viewed Original Estimate Priority Rank Remaining Estimate Reporter Resolution Resolved Security Level	Resolution Details
Add >>	<< Remove

Error message (optional): [   ]

Display a customized error message. A default error message will be displayed if nothing is specified.

Ignore context: ☐

If checked a field will be required by this checker, even if its context is not configured for the current issue.

> ℹ For more information see the JSU Documentation.

Add   Cancel

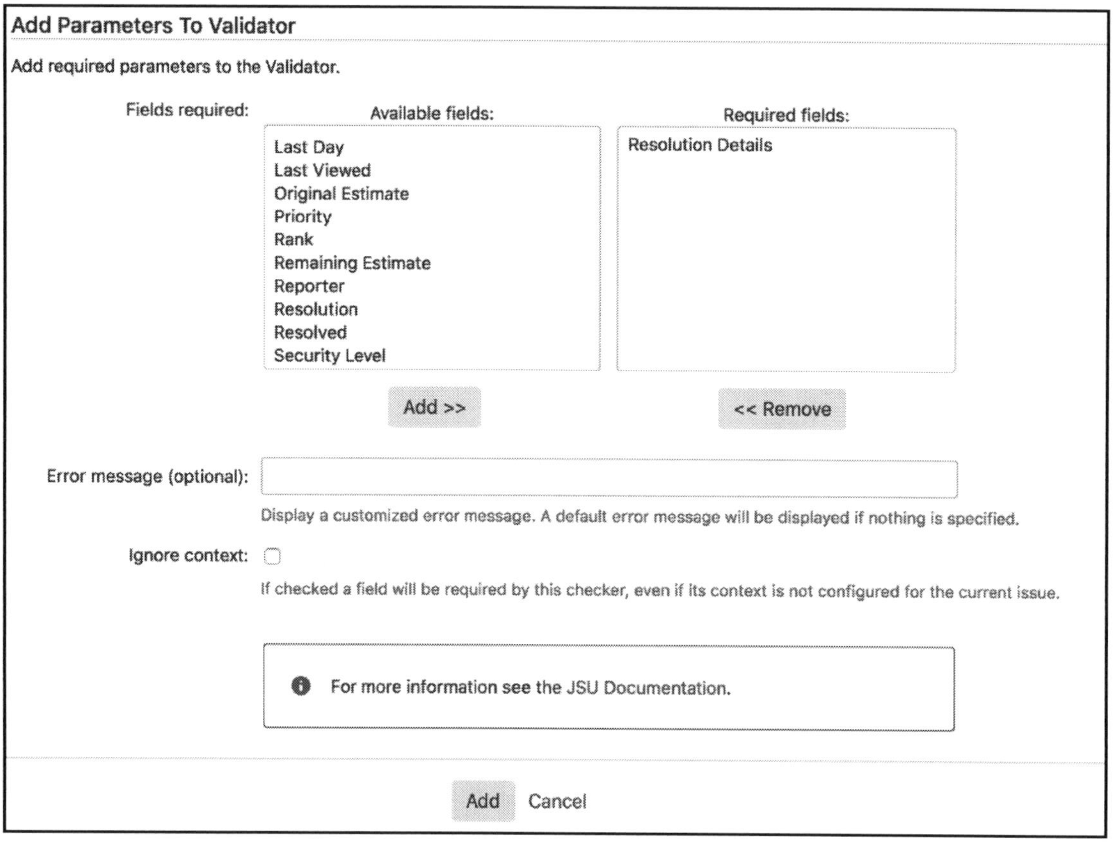

Similar to conditions, when there are multiple validators added to a transition, they form a **validator group**. Unlike conditions, you can only use the logical AND condition for the group. This means that in order to complete a transition, every validator added to the transition must pass its validation criteria. Transitions cannot selectively pass validations using the logical OR condition.

The following screenshot shows a validator (the **Fields required** validator from Suite Utilities for Jira; refer to the *Extending a workflow with workflow add-ons* section) being placed on the transition, which validates whether the user has entered a value for the **Resolution Details** field:

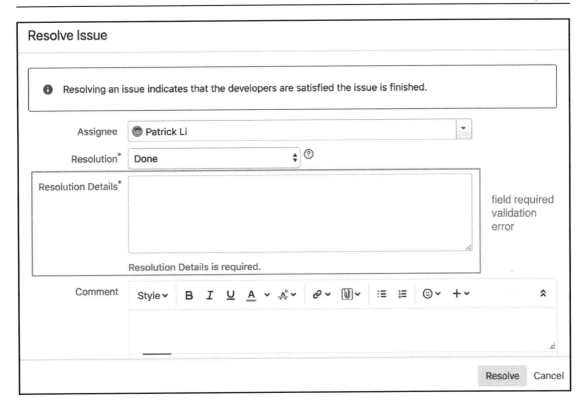

Adding a post function to transitions

Transitions, by default, are created with several post functions. These post functions provide key services to Jira internal operations, so they cannot be deleted from the transition. These post functions perform the following:

- Set the issue status to the linked status of the destination workflow step
- Add a comment to an issue if one is entered during a transition
- Update the change history for an issue and store the issue in the database
- Re-index an issue to keep indexes in sync with the database
- Fire an event that can be processed by the listeners

As you can see, these post functions provide some of the basic functions such as updating a search index and setting an issue's status after transition execution, which is essential in Jira. Therefore, instead of letting users manually add them in and risk the possibility of leaving one or more out, Jira adds them for you automatically when you create a new transition:

1. Select the transition you want to add post functions to.
2. Click on the **Post Functions** link.
3. Click on the **Add post function** link and select the post function you want to add:

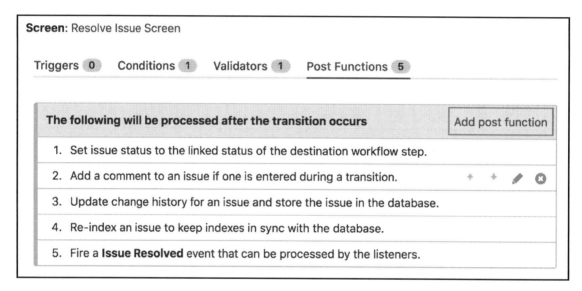

4. Click on the **Add** button to add the post function.
5. Depending on the post function, you may be presented with the **Add Parameters To Function** page where you can specify configuration options for the post function. The following screenshot shows an example from the **Update Issue Field** post function:

Add Parameters To Function

Add required parameters to the Function.

Issue Field: [Assignee   ◇]
The field to change.

Field Value: ⦿ Unassigned
             ○ Automatic
             ○ [   ] ♣
             Start typing to get a list of possible matches.

> ⓘ Please make sure that the value you set is valid for the project using this workflow. Otherwise, the transition may fail at execution time.

[Add] Cancel

When a transition is executed, each post function is executed sequentially as it appears in the list, from top to bottom. If any post function in the list encounters an error during processing, you will receive an error, and the remaining post functions will not be executed.

Since post functions are executed sequentially and some of them possess the ability to modify values and perform other tasks, often, their sequence of execution becomes very important. For example, if you have a post function that changes the issue's assignee to the current user and another post function that updates an issue field's value with the issue's assignee, obviously the update assignee post function needs to occur first, so you need to make sure that it is above the other post function.

You can move the position of post functions up and down along the list by clicking on the **Move Up** and **Move Down** links. Note that not all post functions can be repositioned, such as the re-index issue and fire issue event post functions. They are locked in their positions to ensure data integrity is maintained in Jira.

Updating an existing workflow

Jira lets you make changes to both active and inactive workflows. However, with active workflows, there are several restrictions:

- Existing workflow steps cannot be deleted
- The associated status for an existing step cannot be edited
- If an existing step has no outgoing transitions, it cannot have any new outgoing transitions added

If you need to make these changes, you will have to either deactivate the workflow by removing the associations of the workflow with all projects, or create a copy of the workflow.

You can always make a copy of the active workflow, make your changes, and then swap the original with the copied workflow in your workflow scheme.

When editing an active workflow, you are actually making changes to a draft copy of the workflow created by Jira. All the changes you make will not be applied until you publish your draft.

Do not forget to publish your draft after you have made your changes.

Publishing a draft is a very simple process. All you have to do is as follows:

1. Click on the **Publish Draft** button. You will be prompted if you would like to first create a backup of the original workflow. It is recommended that you create a backup in case you need to undo your changes.
2. Select either **Yes** or **No** to create a backup of the current workflow before applying the changes. This is a handy way to quickly create a backup if you have not made a copy already. If you do choose to create a backup, it is a good idea to name your workflow with a consistent convention (for example, based on a version such as `Sales Workflow 1.0`) to keep track of the changes.
3. Click on the **Publish** button to publish the draft workflow and apply changes, as shown in the following screenshot:

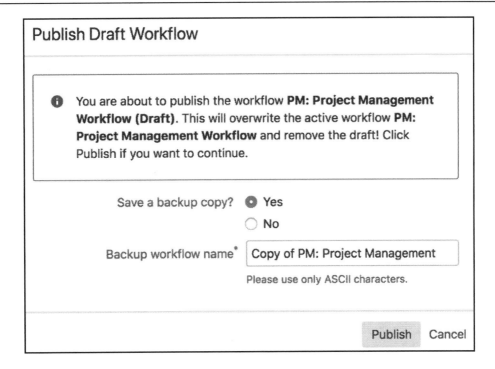

Workflow schemes

While workflows define and model business processes, there still needs to be a way to tell Jira the situations in which to apply the workflows. As with other configurations in Jira, this is achieved through the use of schemes. As we have seen in the previous chapters, schemes act as self-contained, reusable configuration units that associate specific configuration options with projects and, optionally, issue types.

A workflow scheme establishes the association between workflows and issue types. The scheme can then be applied to multiple projects. Once applied, the workflows within the scheme become active.

To view and manage workflow schemes, perform the following steps:

1. Log in as a Jira administrator user.
2. Browse to the Jira administration console.

3. Select the **Issues** tab and then the **Workflow schemes** option. This will bring up the **Workflow schemes** page, as shown in the following screenshot:

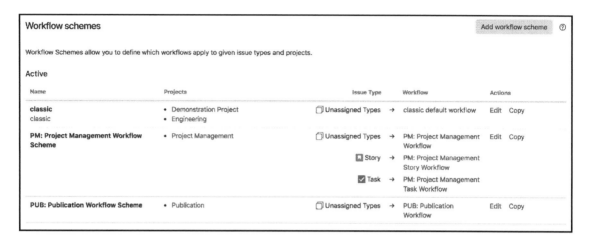

The **Workflow schemes** page shows each scheme's workflow association. For example, in the preceding screenshot, we can see that for **PM: Project Management Workflow Scheme**, the **Task** issue type is assigned with **PM: Project Management Task Workflow**, while the **Story** issue type is assigned to **PM: Project Management Story Workflow**.

Creating a workflow scheme

When a new project is created, a new workflow scheme will be created automatically for the project, so normally, you will not need to create new workflow schemes. However, there might be times, such as when experimenting with changes to the workflow, where you still want to keep existing configurations untouched as a backup. To create a new workflow scheme, perform the following:

1. Browse to the **Workflow schemes** page.
2. Click on the **Add workflow scheme** button. This will take you to the **Add Workflow Scheme** dialog.
3. Enter a name and description for the new workflow scheme. For example, you can choose to name your workflow after the project/issue type it will be applied to.
4. Click on the **Add** button to create the workflow scheme.

You will be taken back to the **Workflow schemes** page once the new scheme has been created, and it will be listed in the table of available workflow schemes.

When you first create a new workflow scheme, the scheme is empty. This means it contains no associations of workflows and issue types, except the default association called Jira Workflow (**jira**). What you need to do next is configure the associations by assigning workflows to issue types.

> You can delete the default Jira Workflow (**jira**) association after you have added an association yourself.

Configuring a workflow scheme

Workflow schemes contain associations between issue types and workflows. After you have created a workflow scheme, you need to configure and maintain the associations as your requirements change. For example, when a new issue type is added to the projects using the workflow scheme, you may need to add an explicit association for the new issue type.

To configure a workflow scheme, perform the following steps:

1. Browse to the **Workflow schemes** page.
2. Click on the **Edit** link for the workflow scheme you want to configure. This will take you to the workflow's details page, as shown in the following screenshot:

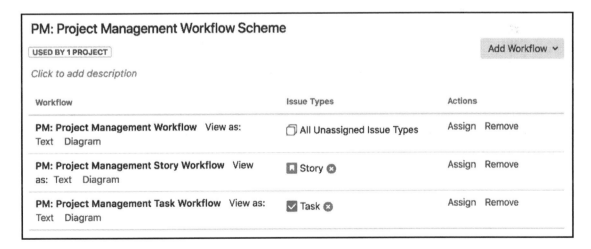

From this page, you will be able to see a list of existing associations, create new associations for issue types, and delete associations that are no longer relevant.

Assigning an issue type to a workflow

Issue types and workflows have a many-to-one relationship. This means each issue type can be associated with one and only one workflow. One workflow can be associated with multiple issue types. This rule is applied on a per-workflow scheme basis, so you can have a different association of the same issue type in a different workflow scheme.

When you add a new association, Jira will list all the issue types and all available workflows. Once you have assigned a workflow to the issue type, it will not appear in the list again until you remove the original association.

Among the list of issue types, there is an option called **All Unassigned Issue Types**. This option acts as a catch-all option for issue types that do not have an explicit association. This is a very handy feature if all issue types in your project are to have the same workflow; instead of mapping them out manually one by one, you can simply assign the workflow to all with this option. This option is also important as new issue types are added and assigned to a project; they will automatically be assigned to the catch-all workflow. If you do not have an **All Unassigned Issue Types** association, new or unassigned issue types will be assigned to use the default basic **jira** workflow. As with normal issue types, you can have only one catch-all association.

 If all issues types will be using the same workflow, use the **All Unassigned Issue Types** option.

There are two ways to assign a workflow to an issue type. If you want to add an issue type to one of the existing associations, perform the following steps:

1. Browse to the workflow scheme's details page for the workflow scheme you want to configure by clicking on its **Edit** link
2. Click on the **Assign** link for the association you want to add an issue type to
3. Select the issue types to add from the **Assign Issue Type to Workflow** dialog
4. Click on the **Finish** button

If you want to create a new association from scratch, perform the following:

1. Browse to the workflow scheme's details page for the workflow scheme you want to configure.
2. Select the **Add Existing** option from the **Add Workflow** menu. This will bring up the **Add Existing Workflow** dialog:

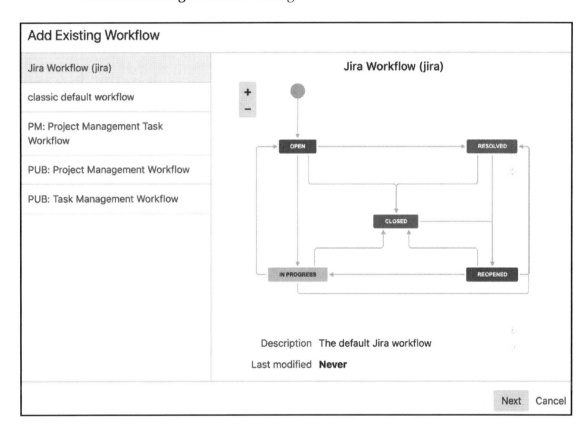

3. Select the workflow to use and click on the **Next** button.

4. Select the issue types to associate with the workflow and click on the **Finish** button. If you select an issue type that is already assigned, it will be removed from the old assignment and added to the currently selected workflow:

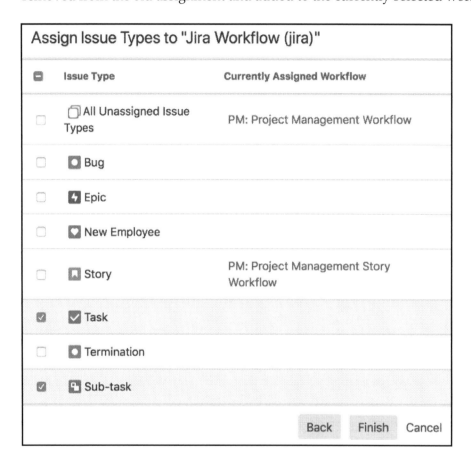

Editing or deleting an association

Once you have associated an issue type to a workflow in a scheme, you cannot add a new association for the same issue type. There is also a **no edit** option to change the association. What you need to do is to delete the existing association and create a new one using the following steps:

1. Browse to the workflow scheme's details page for the workflow scheme you want to configure
2. Click on the **Remove** link for the association you want to remove

Once an association is deleted, you will be able to create a new one for the issue type. If you do not assign a new workflow to the issue type, the workflow with the **All Unassigned Issue Types** option will be applied.

Applying a workflow scheme to projects

Workflow schemes are inactive by default after they are created. This means there are no projects in Jira using the workflow scheme. To activate a workflow scheme, you need to select the scheme and apply it to the project.

When assigning a workflow scheme to a project, you need to follow the four basic steps:

1. Browse to the project administration page for the project you want to apply the workflow scheme to
2. Select the **Workflows** option from the left panel
3. Click on the **Switch Scheme** button
4. Select the new workflow scheme to use and click on the **Associate** button

On the confirmation page, depending on the differences between the current and new workflow, you will be prompted to make migration decisions for existing issues. For example, if the current workflow has a status called **Resolved** and the new workflow does not (or it has something equivalent but with a different ID), you need to specify the new status to place the issues that are currently in the **Resolved** status. Once mapped, Jira will start migrating existing issues to the new status:

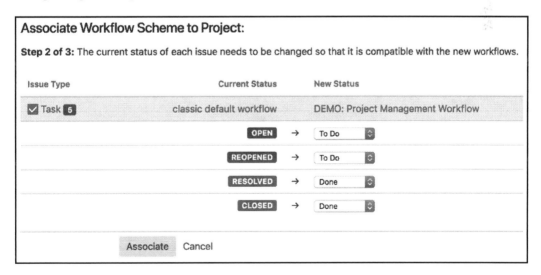

Given below are the steps:

1. Select new workflow statuses for the existing issues that are in statuses that do not exist in the new workflow
2. Click on the **Associate** button to start the migration

Once the migration starts, Jira will display a progress bar showing you the progress. Depending on the number of issues that need to be migrated, this process may take some time. It is recommended to allocate a time frame to perform this task as it can be quite resource-intensive for large instances.

Delegated workflow management

Just like we have seen in Chapter 6, *Screen Management*, project administrators have also been empowered to make changes to workflows that are used only by their own projects, instead of having to rely entirely on the Jira administrator.

There are, however, some restrictions to this:

- Only existing statuses can be used in the workflow
- If the status is already used by an issue in the project, the status cannot be deleted from the workflow
- Transition properties, conditions, validators, and post functions cannot be updated
- The workflow must not be shared with any other projects
- Workflows can only be updated in diagram mode

This allows you to make changes such as adjusting statuses and transition flow to workflows that are dedicated to a single project, which would normally be the case since new workflows are automatically created with each new project. To make changes to workflows for your project as a project administrator, perform the following steps:

1. Browse to the target project's administration page.
2. Click on the **Workflows** option from the left panel.
3. Select the workflow you want to edit and click the **Edit** button.
4. Click the **Publish** button once you are done with your changes:

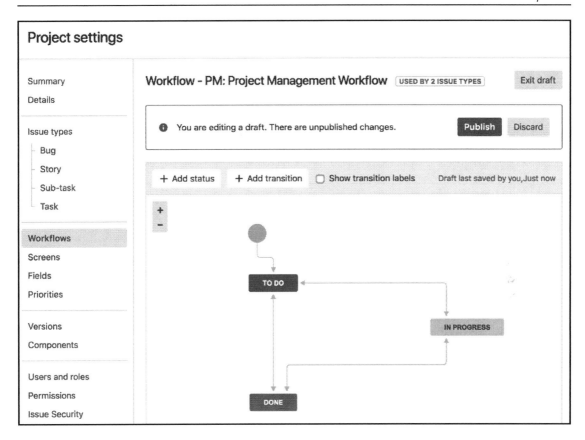

Extending a workflow with workflow add-ons

There are a number of very useful add-ons that will provide additional components such as conditions, validators, and post functions. The following presents some of the most popular workflow-related plugins.

Suite utilities for Jira

You can find a number of very useful conditions, validators, and post functions with this add-on. For example, the **Update Issue Field** post function that ships with Jira allows you to update any issue fields such as priority and assignee when a workflow transition completes. The Suite Utilities for Jira add-on complements this by providing a very similar **Update Any Issue Field** post function, which handles custom fields. There are many other useful components such as the **Copy Value From Other Field** post function, which will allow you to implement some amazing logic with your workflow. It is a must-have add-on for any Jira. You can find out more at https://marketplace.atlassian.com/apps/5048/jsu-suite-utilities-for-jira.

Jira workflow toolbox

As the name suggests, it's a workflow toolbox with a rich set of workflow conditions, validators, and post functions intended to fill many gaps when developing complex workflows. For example, it provides a condition and validator that allows you to specify the checking rules with regular expressions. You can find out more at https://marketplace.atlassian.com/apps/29496/jira-workflow-toolbox.

Jira Misc workflow extensions

This is another plugin with an assortment of conditions, validators, and post functions. Normal post functions let you alter the current issue's field values. This plugin provides post functions that will allow you to set a parent issue's field values from subtasks along with many other features. You can find out more at https://marketplace.atlassian.com/apps/292/jira-misc-workflow-extensions.

Workflow enhancer for Jira

This contains a variety of validators and conditions around comparisons of the value of a field with another field, and lets you set up validation logic to compare dates, numeric, and Boolean values; you can find out more at https://marketplace.atlassian.com/apps/575829/workflow-enhancer-for-jira.

ScriptRunner for Jira

This is a very useful and powerful add-on that allows you to create your own custom conditions, validators, and post functions by writing scripts. This does require you to have some programming knowledge and a good understanding of Jira's API. You can find out more at `https://marketplace.atlassian.com/apps/6820/scriptrunner-for-jira`.

The HR project

We have seen the power of workflows and how we can enhance the usefulness of Jira by adapting to everyday business processes. With our HR project, we have already defined two issue types to represent the onboarding and dismissal of an employee; both of these use the same default workflow with two steps, *to do* and *done*. So, we will now customize the workflow to represent a real-world HR process.

Our requirements for the business process would then include the following:

- The `New Employee` and `Termination` issue types will use a customized workflow, while the **Task** issue type will continue to use the existing one
- For the `Termination` issue type, we will add two additional steps, one to conduct an exit interview, and one to ensure that all necessary company assets are returned
- Ensure that only authorized personnel can transition the issue through the various statuses of the workflow

The easiest way to implement these requirements would be to create a new workflow and add the additional process steps as new statuses. We will first do this to get our workflow structure in place. Later on, we will also look at how we can use other features in Jira and incorporate them into our workflow to make it more robust.

Setting up workflows

The first step is to create a new workflow for our `Termination` issue type, since we still want to keep the existing workflow for the **Task** issue type. The easiest way to get started is to clone the current workflow to save us some time:

1. Browse to the **View Workflows** page
2. Click on the **Copy** link for the `HR: Task Management Workflow` workflow

3. Name the new workflow `HR: Termination Workflow`

4. Click on the **Copy** button to create our workflow

The next step is to add in the extra status we need. Make sure that you are in the workflow designer by selecting the **Diagram** option:

1. Click on the **Add status** button.

2. Enter the name for our new status as `In Exit Review`, set the **Category** to **In Progress**, and click on **Add**. You will need to hit the *Enter* key on your keyboard since we are creating a new status.

3. Click on the **Create** button to create the workflow status.

4. Repeat steps 2 and 3 to create a new status called `Collecting Assets`.

Now that we have our statuses added to our workflow, we need to connect them into the workflow with transitions. For now, we will make the workflow to go in a sequence in the order of **To Do** | **In Exit Review** | **Collecting Assets** | **Done**. Let's start with creating a transition going from **To Do** | **In Exit Review**:

1. Click on the **Add transition** button

2. Select **To Do** as the **From status**

3. Select **In Exit Review** as the **To status**

4. Name the new transition as `Conduct Exit Review`

5. Select **Workflow Screen** for **Screen**

6. Click on the **Add** button to create the transition

7. Repeat steps 1 to 6 to create two more transitions, linking **In Exit Review** to **Collecting Assets**, and **Collecting Assets** to **Done**

8. With the new transitions in place, we will also want to remove the existing transitions between **To Do** and **Done**, so people cannot skip the process steps

Your workflow will look something like the one shown in the following screenshot. You can rearrange the elements in the workflow to make the diagram flow more naturally:

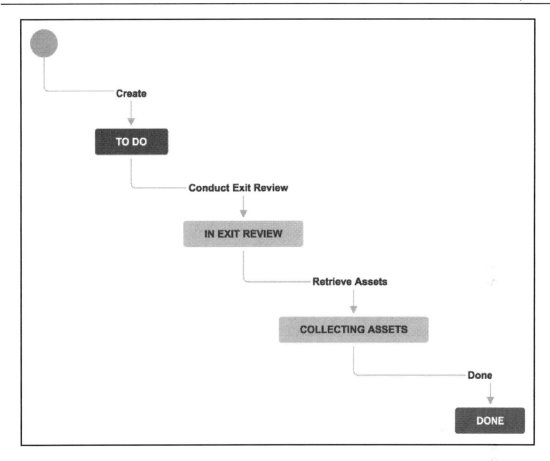

The next customization we will do is to make sure that only authorized personnel can transition the issue along the workflow. For now, we will set it so only members of the **jira-administrators** group can transition an issue after it is created. Once we cover Chapter 9, *Securing Jira*, we can change this security setting:

1. Click on the Conduct Exit Review transition and click on **Conditions** from the *transition property* section
2. Click on the **Add condition** button to bring up the **Add Condition To Transition** page
3. Select the **User Is In Group** option
4. Select the **jira-administrator** group
5. Click on **Add** to add the condition to the transition
6. Repeat steps 1 to 5 on the remaining transitions

Using the **Users Is In Group** option will ensure that only users in the selected group, `jira-administrator` in this case, will see the transition with the condition applied to it.

Applying the workflow

With our workflow in place and set up, we need to let Jira know the issue types that will be using our new workflow. Since we already have a workflow scheme in place for our project, we just need to associate the appropriate issue type to the workflow:

1. Browse to the **Workflow schemes** page
2. Click on the **Edit** link for `HR: Task Management Workflow Scheme`
3. Click on the **Add Workflow** menu and select the **Add Existing** option
4. Select our new `HR: Termination Workflow` option and click on the **Next** button
5. Select the `Termination` issue type
6. Click on **Finish** to create the association
7. Click on the **Publish** button to apply the change

This associates our new workflow with the `Termination` issue type specifically created for our `HR` project, and leaves the default workflow for the others.

Putting it together

With our new workflow in place, we can now create a new `Termination` issue and start testing our implementation. Since we need to simulate a scenario where an *unauthorized user* cannot transition the issue after it is created, we need to create a new user. We will look at user management and security in `Chapter 9`, *Securing Jira*. For now, we will simply add a new user to our system:

1. Browse to the Jira administration console
2. Select the **Users management** tab and click on the **Users** link
3. Click on the **Create user** button to bring up the **Create New User** dialog
4. Name the new user `john.doe` (John Doe)
5. Set the password and email address for this new user
6. Uncheck the **Send Notification Email** option
7. Check the `Jira Software` option for **Application access**
8. Click on the **Create** button to create the user

Now, log in to Jira as a new business user, john.doe, and create a new termination issue. After you create the issue, you will notice that you cannot execute any transitions. This is because you (john.doe) are not in the **jira-administrators** group. The administrator user we created in Chapter 1, *Getting Started with Jira*, is in the **jira-administrators** group, so let's log in as the administrator. Once logged in as the administrator, you will see our new transition, **Conduct Exit Interview**, as shown in the following screenshot:

You will also see that, if you create a new task in the HR project, the task issue will continue to use the default workflow.

With the current workflow set up, everything happens in sequential order. However, sometimes, you might need things to happen in parallel. For example, in the collecting assets step, there might be multiple assets to be collected for various teams, such as a laptop for IT and a key card for security. It will be a lot more efficient if you can perform them at the same time and be able to track them individually. One way you can do this is by creating subtasks for each asset under the issue (remember, an issue can only be assigned to one person), and assign the subtask to the relevant team such as IT and security, so they can chase up with the employee to retrieve the asset. You can then set a condition on the **Done** transition to make sure that all subtasks are completed before they can be executed.

This can be expanded upon to have the asset collection and exit interview as subtasks so that both can happen at the same time, and you can create different subtask issue types to differentiate them, as covered in `Chapter 4`, *Issue Management*. Your termination issue may look something like this:

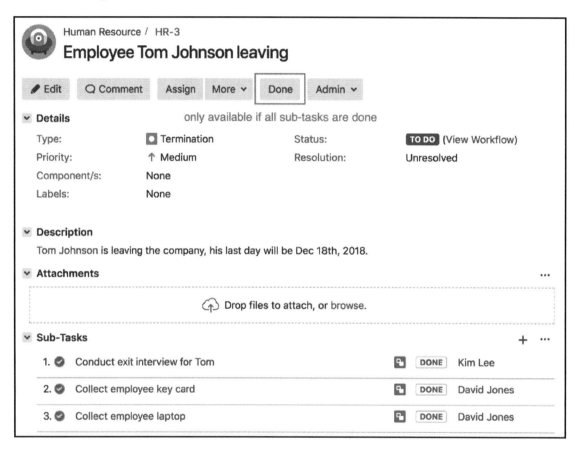

Summary

In this chapter, we looked at how Jira can be customized to adapt to your organization. At the heart of this powerful feature is a robust workflow system that allows you to model Jira workflows based on existing business processes. We also looked at the various components within a workflow, how to perform validations, and how post-processing provides a level of process automation.

In the next chapter, we will look at how we can combine the power of a workflow and its event-driven systems to facilitate communication through Jira notifications and the email system.

Emails and Notifications

8

So far, you have learned how to use and interact with Jira directly from its web interface through a browser. We will now take a look at how Jira uses emails as a notification mechanism to alert you of updates.

One powerful feature of Jira is its ability to create new issues, add comments to issues, and update issue details through emails. This provides you with a whole new option of how you and your users can interact with Jira.

By the end of this chapter, you will have learned about the following topics:

- How to set up a mail server in Jira
- Events and how they are related to notifications
- How to configure Jira to send out notifications based on events
- How to create custom mail templates
- What a mail handler is
- How to create issues and comments by sending emails to Jira

Jira and email

Emails have become one of the most ubiquitous communication tools in today's world. Businesses and individuals rely on emails to send and receive information around the world almost instantly. Therefore, it should come as no surprise that Jira is fully equipped and integrated with email support.

Jira's email support comes in several flavors. First, Jira sends emails to users about changes that have been made to their issues, such as adding comments, so people working on the same issue can be kept on the same page. Second, Jira can also poll mailboxes for emails and create issues and comments based on the content of the email. The last feature is the ability for users to create and subscribe to filters to set up feeds in Jira (we will discuss filters in Chapter 10, *Searching, Reporting, and Analysis*). These features open up a whole new dimension on how users can interact with Jira.

In the following sections, we will look at what you need to do to enable Jira's powerful email support and also explore the tools and options at your disposal so that you can configure Jira and email your own way. The following diagram shows how Jira interacts with various mail servers:

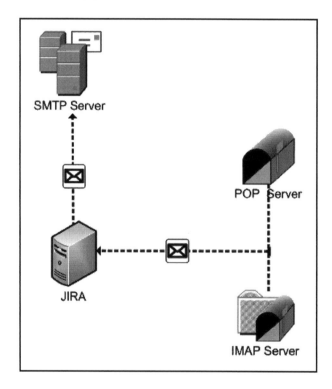

Mail servers

For Jira to communicate with emails, you need to configure or register your mail servers in Jira. There are two types of mail servers you need to configure:

- **Outgoing**: This is used by Jira to send emails out to users. Jira supports SMTP mail servers.
- **Incoming**: This is used by Jira to retrieve emails from users. Jira supports POP and IMAP servers.

We will start with outgoing mail servers to see how we can configure Jira to send emails to users as well as customize the email's contents.

Working with outgoing mail

Like many settings in Jira, you need to be a Jira system administrator (the user that's created during the initial setup is a system administrator) to configure mail server details. Perform the following steps to manage the outgoing mail server:

1. Log in to Jira as a Jira system administrator.
2. Browse to the Jira administration console.
3. Select the **System** tab and then the **Outgoing Mail** option. This will bring up the **Outgoing Mail** page:

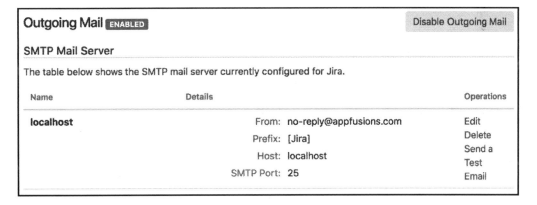

 You can have only one outgoing mail server in Jira.

Adding an outgoing mail server

There are two ways to add an outgoing mail server in Jira. Both options have some common configuration parameters that you will need to fill in. The following table shows these parameters:

Field	Description
Name	This specifies a name for the mail server.
Description	This specifies a brief description of the mail server.
From address	This specifies an email address that outgoing emails will appear to have come from.
Email prefix	This specifies a prefix that will appear with all the emails sent from Jira. This allows your users to set up filter rules in their mail clients. The prefix will be added to the beginning of the email subject.
Service Provider	Select from one of the three predefined mail providers, that is, Google, Yahoo!, or the custom SMTP server.
Host Name	This specifies the hostname of your mail server (for example, `smtp.example.com`).
SMTP Port	This specifies the port number that your mail server will be running on. This is optional; if left blank, the default port number 25 will be used.
Username	This is used to authenticate against the mail server if required. Note that mail servers may require authentication to relay emails to non-local users.
Password	This is used to authenticate the user against the mail server if required.
JNDI Location	This is the JNDI lookup name if you already have a mail server configured for your application server. Refer to the following section for details.

For the rest of the parameters, depending on which option you select to set up your mail server, you only need to fill in the ones that are appropriate.

The first option is to select from one of the built-in service providers and specify the mail server's details. For example, if you have an SMTP mail server running, you can select the **Custom** option from the **Service Provider** field and specify the host and port number. This is the approach most people will use as it is simple and straightforward. With this approach, the administrator fills in the mail server's host information, such as the hostname and port number:

1. Browse to the **Outgoing Mail** page.
2. Click on the **Configure new SMTP mail server** button.

3. Enter the general details of your mail server, including the name, description, from address, and email prefix.
4. Select the type of mail server from the **Service Provider** field.
5. Enter the mail server's connection details.
6. Click on the **Test Connection** button to verify the configuration.
7. Click on the **Add** button to register to the mail server:

Add SMTP Mail Server

Use this page to add a new SMTP mail server. This server will be used to send all outgoing mail from Jira.

Name

The name of this server within Jira.

Description

From address

The default address this server will use to send emails from.

Email prefix

This prefix will be prepended to all outgoing email subjects.

Server Details
Enter *either* the host name of your SMTP server *or* the JNDI location of a javax.mail.Session object to use.

SMTP Host

Service Provider   `Custom`

Protocol   `SMTP`

Host Name

The SMTP host name of your mail server.

SMTP Port

Optional – SMTP port number to use. Leave blank for default (defaults: SMTP - 25, SMTPS - 465).

Timeout   `10000`

Timeout in milliseconds - 0 or negative values indicate infinite timeout. Leave blank for default (10000 ms).

TLS.   ☐

Optional – the mail server requires the use of TLS security.

Username

Optional – if you use authenticated SMTP to send email, enter your username.

Password

Optional – as above, enter your password if you use authenticated SMTP.

or

JNDI Location

JNDI Location

The JNDI location of a javax.mail.Session object, which has already been set up in Jira's application server.

Test Connection   Add   Cancel

Jira comes with support for Google and Yahoo! mail services. You can select these options in the **Service Provider** field if you are using these services.

The second option is to use **JNDI**. This approach is slightly more complicated as it requires configuration on the application server itself, and it requires you to restart Jira.

If you are using the standalone distribution, which uses Apache Tomcat, the JNDI location will be `java:comp/env/mail/JiraMailServer`. You will also need to specify the mail server details as a JNDI resource in the `server.xml` file in previous line `JIRA_INSTALL/conf` directory.

A sample declaration for Apache Tomcat is shown in the following code snippet. You will need to substitute some values with the real values for some of the parameters in the code of your mail server's details:

```
<Resource name="mail/JiraMailServer" auth="Container"
type="javax.mail.Session"
  mail.smtp.host="mail.server.host"
  mail.smtp.port="25"
  mail.transport.protocol="smtp"
  mail.smtp.auth="true"
  mail.smtp.user="username"
  password="password"
/>
```

You will need to restart Jira after saving your changes to the `server.xml` file.

Disabling outgoing mail

If you are running a test or evaluation Jira instance, or testing changes that you have made to your configurations, you might not want to flood your users with test emails. The easiest way for you to disable all outgoing emails is by just clicking on the **Disable Outgoing Mail** button. This will stop Jira from sending emails as a result of issue updates. Once you are ready to send emails again, you can click on the **Enable Outgoing Mail** button.

Disabling outgoing mail will only prevent Jira from sending out notification emails based on notification schemes.

Enabling SMTP over SSL

To increase security, you can encrypt the communication between Jira and your mail server if your mail server supports SSL. There are two steps involved in enabling SSL over SMTP in Jira:

- The first step is to import your mail server's SSL certificate into Java's trust store. You can do this with Java's keytool utility. On a Windows machine, run the following command in Command Prompt:

```
Keytool -import -alias mail.yourcompany.com -keystore
$JAVA_HOME/jre/lib/security/cacerts -file yourcertificate
```

- The second step is to configure your application server to use SSL for mail communication. The following declaration is for Apache Tomcat, which is used by Jira standalone. We use the same configuration file and only need to add two additional parameters:

```
<Resource name="mail/JiraMailServer"
  auth="Container"
  type="javax.mail.Session"
  mail.smtp.host="mail.server.host"
  mail.smtp.port="25"
  mail.transport.protocol="smtp"
  mail.smtp.auth="true"
  mail.smtp.user="username"
  password="password"
  mail.smtp.atarttls.enabled="true"
mail.smtp.socketFactory.class="javax.net.ssl.SSLSocketFactory"
  />
```

Once you have imported your certificate and configured your mail server, you will have to restart Jira.

Sending a test email

It is always a good idea to send a test email after you configure your SMTP mail server to make sure that the server is running and that you have set it correctly in Jira:

1. Browse to the **Outgoing Mail** page.
2. Click on the **Send a Test Email** link for your SMTP mail server.
3. Click on the **Send** button to send the email. Jira will autofill the **To** address based on the user you have logged in as.

If everything is correct, you will see a confirmation message in the **Mail log** section and receive the email in your inbox. If there are errors, such as mail server connection, then the **Mail log** section will display these problems. This is very useful when troubleshooting any problems with Jira's connectivity with the SMTP server:

Send email

You can send a test email here.

To	patrick@appfusions.com
Subject	Test Message From JIRA
Message Type	Text
Body	This is a test message from JIRA. Server: Default SMTP Server SMTP Port: 25 Description: From: no-reply@appfusions.com Host User Name: null
SMTP logging	☐ Log SMTP-level details

[Send]   Cancel

Mail log

Log
```
An error has occurred with sending the test email:
com.atlassian.mail.MailException: com.sun.mail.util.MailConnectException: Couldn't connect to host, port:
localhost, 25; timeout 10000;
  nested exception is:
    java.net.ConnectException: Connection refused
    at com.atlassian.mail.server.impl.SMTPMailServerImpl.sendWithMessageId(SMTPMailServerImpl.java:222)
    at com.atlassian.mail.server.impl.SMTPMailServerImpl.send(SMTPMailServerImpl.java:159)
    at com.atlassian.jira.plugins.mail.webwork.SendTestMail.doExecute(SendTestMail.java:95)
    at webwork.action.ActionSupport.execute(ActionSupport.java:165)
    at com.atlassian.jira.action..JiraActionSupport.execute(JiraActionSupport.java:63)
    at webwork.interceptor.DefaultInterceptorChain.proceed(DefaultInterceptorChain.java:39)
    at webwork.interceptor.NestedInterceptorChain.proceed(NestedInterceptorChain.java:31)
    at webwork.interceptor.ChainedInterceptor.intercept(ChainedInterceptor.java:16)
    at webwork.interceptor.DefaultInterceptorChain.proceed(DefaultInterceptorChain.java:35)
    at webwork.dispatcher.GenericDispatcher.executeAction(GenericDispatcher.java:225)
```
Log of the events for sending mail.

In the preceding screenshot, you can see that test email delivery has failed, and the error is because Jira was unable to connect to the configured SMTP server.

Mail queues

Emails in Jira are not sent immediately when an operation is performed. Instead, they are placed in a mail queue, which Jira empties periodically (every minute). This is very similar to the real-life scenario, where mail is placed in mailboxes and picked up every day by postal workers.

Viewing the mail queue

Normally, you do not need to manage the mail queue. Jira automatically places emails into the queue and flushes them periodically. However, as an administrator, there may be times when you wish to inspect the mail queue, especially to troubleshoot problems related to Jira notification emails. Sometimes, emails can get stuck for a number of reasons, and inspecting the mail queue will help you identify these problems and fix them.

Perform the following steps to view the content of the mail queue:

1. Browse to the Jira administration console.
2. Select the **System** tab and then the **Mail queue** option:

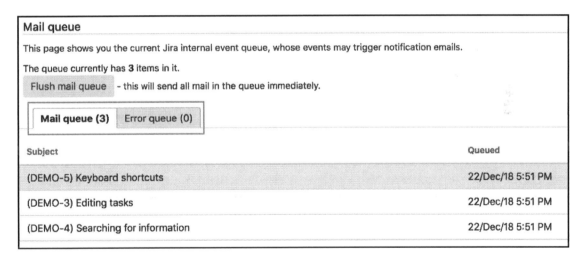

This page provides you with a one-page view of the current emails in the queue that are waiting for delivery. There are two queues—the main mail queue and the error queue.

The **Mail queue** tab contains all the emails that are pending to be delivered. If Jira is able to successfully deliver these emails, they will be removed from the queue. Items listed in red indicate that Jira has unsuccessfully attempted to send those emails. Jira will retry 10 times; if still unsuccessful, these items will be moved to the error queue.

The **Error queue** tab contains emails that cannot be delivered by Jira. You can choose to resend all the failed items in the error queue or delete them.

Flushing the mail queue

While Jira automatically flushes the mail queue, you can also manually flush the queue if it gets stuck or sends out emails immediately. When you manually flush the queue, Jira will try to send all the emails that are currently in the queue.

Perform the following steps to manually flush the mail queue:

1. Browse to the **Mail queue** page
2. Click on the **Flush mail queue** button

If Jira is successful in sending emails, you will see the queue shrink and the items disappear. If some emails fail to be delivered, those items will be highlighted in red.

Manually sending emails

Sometimes, you, as the administrator, may need to send out emails containing important messages to a wide audience. For example, if you are planning some maintenance work that will take Jira offline for an extended period of time, you may want to send an email to all Jira users to let them know of the outage.

Jira has a built-in facility where you can manually send out emails to specific groups of users. There are two options when manually sending emails—you can send them groups or by projects.

When sending by groups, all you have to do is select one or more groups in Jira, and all users that belong to the selected groups will receive the email. Users belonging to more than one group will not get duplicated emails.

When sending emails by projects, you have to first select one or more projects and then the project roles. We will discuss project roles in more detail in the next chapter; for now, you can think of them as groups of users within projects. For example, you can send emails to all users that are part of the demonstration project rather than all users in Jira.

To send emails to users in Jira, perform the following steps:

1. Browse to the Jira administration console
2. Select the **System** tab and then the **Send email** option
3. Choose if you want to send to users by **Project Roles** or **Groups**
4. Enter the email's **Subject** and **Body** content
5. Click on the **Send** button to send the email to all users in the selected project roles/groups

The following screenshot shows an example of sending maintenance outage notification emails to everyone by selecting the **jira-software-users** group, which every Jira Software user is a member of by default:

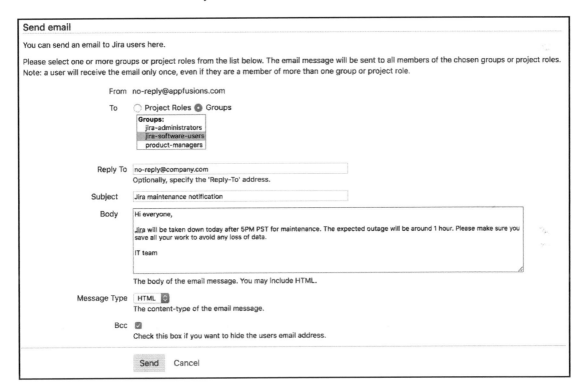

 Since Jira does not provide a **What You See Is What You Get (WYSIWYG)** editor for composing emails, you may want to draft an email and send it to yourself before sending it out to everyone.

Events

Jira is an event-driven system. This means that, when an action occurs (for example, when an issue is created), Jira usually fires off a corresponding event. This event is then picked up by components that are designed to listen to the event. Not surprisingly, these are called **listeners**. When a listener picks up an event, it will perform its duty, such as keeping issues up to date with changes or sending an email to users who are watching the issue.

This mechanism allows Jira to process operations asynchronously. The advantage of this model is operations, such as sending emails, and it's separated from Jira's core functions such as issue creation. If there is a problem with the mail server, for example, you will not want this problem to prevent your users from creating issues.

There are two types of event in Jira:

- **System events**: These are internal events that are used by Jira, and they usually represent the main functionalities in Jira. They cannot be added, edited, or deleted.
- **Custom events**: These are events that are created by users. They can be added and deleted, and they are fired through workflow post functions.

The following table lists all the system events in Jira and what they are used for:

Event	Description
Issue Created	This states that an issue has been created in Jira.
Issue Updated	This states that an issue has been updated (for example, changes to its fields).
Issue Assigned	This states that an issue has been assigned to a user.
Issue Resolved	This states that an issue has been resolved (usually applied to the resolve workflow transition).
Issue Closed	This states that an issue has been closed (usually applied to the closed workflow transition).
Issue Commented	This states that a comment has been added to an issue.
Issue Comment Edited	This states that a comment has been updated.
Issue Reopened	This states that an issue has been reopened (usually applied to the reopen workflow transition).

Issue Deleted	This states that an issue has been deleted from Jira.
Issue Moved	This states that an issue has been moved (to a different or the same project).
Work Logged On Issue	This states that the time has been logged on this issue (if time tracking has been enabled).
Work Started On Issue	This states that the assignee has started working on this issue (usually applied to the start progress workflow transition).
Work Stopped On Issue	This states that the assignee has stopped working on this issue (usually applied to the stop progress workflow transition).
Issue Worklog Updated	This states that the worklog has been updated (if time tracking has been enabled).
Issue Worklog Deleted	This states that the worklog has been deleted (if time tracking has been enabled).
Generic Event	This states that a generic event can be used by any workflow post function.
Custom Event	This states that the events that have created by the user to represent arbitrary events have been generated by business processes.

As an administrator, you will be able to get a one-page view of all the events in Jira. You just need to do the following:

1. Browse to the Jira administration console.
2. Select the **System** tab and then the **Events** option. This will bring up the **View Events** page.

Each event is associated with a template, which is often referred to as a mail template. These templates define the content structure of emails when notifications are sent. For system events, you cannot change their templates (you can change the template files, however). For custom events, you can choose to use one of the existing templates or create your own mail template.

In the following sections, we will first look at how to create and register custom mail templates, create a new custom event to use the new template, and fire the new event when actions are performed on an issue. After that, we will look at how to tie events to notifications so that we can tell Jira who should receive notification emails for the event.

Adding a mail template

Mail templates are physical files that you create and edit directly via a text editor; you cannot edit mail templates in the browser. Each mail template is made up of three files:

- **Subject template**: This file contains the template that's used to generate the email's subject
- **Text template**: This file contains the template that's used by Jira when the email is sent as plain text
- **HTML template**: This file contains the template that's used by Jira when the email is sent as HTML

Mail templates are stored in the `<JIRA_INSTALL>/atlassian-jira/WEB-INF/classes/templates/email` directory. Each of the three files that we listed is placed in its respective directory—`subject`, `text`, and `html`.

While creating new mail templates, it is a good practice to name your template files after the issue event. This will help future users understand the purpose of those templates.

Mail templates use Apache's Velocity template language (`http://velocity.apache.org`). For this reason, creating new mail templates will require some understanding of HTML and template programming.

If your templates only contain static text, you can simply use standard HTML tags for your template. However, if you need to have dynamic data rendered as part of your templates, such as the issue key or summary, you will need to use the **Velocity syntax**. A full explanation of Velocity is beyond the scope of this book. The following section provides a quick introduction to creating simple mail templates for Jira. You can find more information on Velocity and its usage in Jira mail templates at `https://confluence.atlassian.com/x/dQISCw`.

In a Velocity template, all the text will be treated as normal. Anything that starts with a dollar sign (`$`), such as `$issue`, is a Velocity statement. The `$` sign tells Velocity to reference the item after the sign, and, when combined with the period (`.`), you are able to retrieve the value specified. For example, the following code in a template will get the issue key and summary from the current issue, separated by a – character:

```
$issue.key - $issue.summary
```

This would produce content similar to `DEMO-1 - This is a demo issue.`

Jira provides a range of Velocity references that you can use for creating mail templates. These references allow you to access data such as the issue being updated and the user triggering the event. You can find a comprehensive list at `https://developer.atlassian.com/server/jira/platform/jira-templates-and-jsps/#email-templates`.

Now that you have a basic understanding of how Velocity works, you need to create a template for the mail subject. The following code shows a typical subject template:

```
$eventTypeName: ($issue.key) $issue.summary
```

When the template is processed, Jira will substitute the actual values for the event type (for example, issue created), issue key, and issue summary. Therefore, the preceding example would produce content similar to `Issue Escalated: HD-11: Database server is running very slow.`

You then need to create a template for the actual email content. You need to create a text and HTML version. The following code shows a simple example of a text-based template, which displays the key for the escalated issue:

```
Hello,

The ticket $issue.key has been escalated and is currently being worked on.
We will contact you if we require more information.

Regards
Support team.
```

Before Jira sends out the email, the preceding text will be processed, where all Velocity references, such as `$issue.key`, will be converted into proper values, for example, `DEMO-1`.

After creating your mail templates, you need to register them with Jira. To register your new templates, locate and open the `email-templates-id-mappings.xml` file in the `<JIRA_INSTALL>/atlassian-jira/WEB-INF/classes` directory in a text editor. Add a new entry to the end of the file before closing the `</templatemappings>` tag, as follows:

```
<templatemapping id="10001">
  <name>Example Custom Event</name>
  <template>examplecustomevent.vm</template>
  <templatetype>issueevent</templatetype>
</templatemapping>
```

Here, we register a new custom mail template entry. The details of this are given in the following table:

Parameter	Description
id	This is the unique ID for the template. You need to make sure that no other template mapping has the same ID.
name	This is a human-readable name for JIRA to display.
template	These are the mail template filenames for subject, text, and HTML. All three template files must be named as specified here.
type	This is the template type. For events that are generated from an issue, the value will be issueevent.

After creating your templates and registering them in the mapping file, you will have to restart Jira for the changes to be picked up. The new templates will be available when we create new events, which we will talk about in the following section.

Adding a custom event

Jira comes with a comprehensive list of system events that are focused around issue-related operations. However, there will be times when you will need to create custom-designed events that represent specialized business operations, or when you simply need to use a custom email template.

Perform the following steps to add a new custom event:

1. Browse to the **View Events** page.
2. Enter a name and description for the new event in the **Add New Event** section.
3. Select the mail template for the new event.
4. Click on the **Add** button to create a new event:

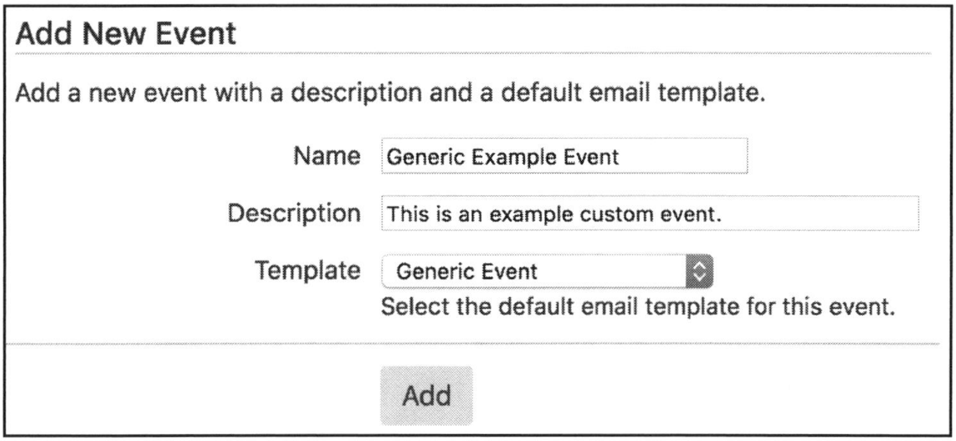

Add New Event

Add a new event with a description and a default email template.

Name | Generic Example Event

Description | This is an example custom event.

Template | Generic Event ◊

Select the default email template for this event.

Add

Firing a custom event

Unlike system events, with custom events, you need to tell Jira when it can fire a custom event.

Custom events are mostly fired by workflow transitions. Recall from Chapter 7, *Workflow and Business Process*, that you can add post functions to workflow transitions. Almost all of Jira's transitions will have a post function that fires an appropriate event. It is important to understand that just because an event is fired does not mean that there needs to be something to listen to it.

If you skipped Chapter 7, *Workflow and Business Process*, or still do not have a good understanding of workflows, now would be a good time to go back and revisit that chapter.

Perform the following steps to fire a custom event from a workflow post function:

1. Browse to the **View Workflows** page.
2. Click on the **Edit** link for the workflow that will be used to fire the event.
3. Click on the transition that will fire the event when executed.
4. Click on the **Post Functions** tab.

5. Click on the **Edit** link for the post function that reads **Fire a <event name> event that can be processed by the listeners**:

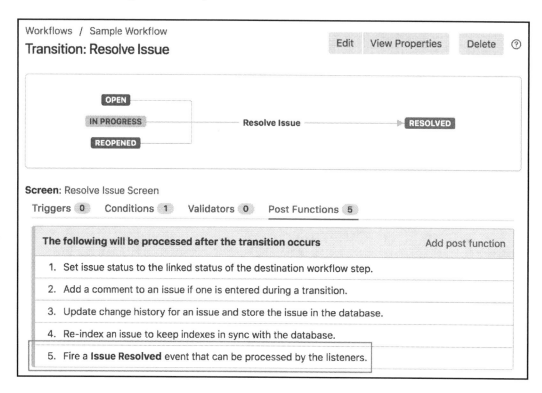

6. Select the custom event from the drop-down list.
7. Click on the **Update** button to apply any changes to the post function.
8. Publish the workflow.

Now, whenever the workflow transition is executed, the post function will run and fire the selected event. Each transition can fire only one event, so you cannot have both **Issue Created** and **Issue Updated** events being fired from the same transition.

Notifications

Notifications associate events (both system and custom) to email recipients. When an event is fired and picked up, emails will be sent out. Notification types define the recipients of emails. For example, you can set them to only send emails to a specific user or all members from a given user group. You can add multiple notifications to a given event.

Jira ships with a comprehensive list of notification types (that is, the recipients) that will cover many of your needs. The following table lists all the notification types that are available and how they work:

Notification type	Description
Current Assignee	This is the current assignee of the issue.
Reporter	This is the reporter of the issue (usually the person who originally created the issue).
Current User	This is the user who fired the event.
Project Lead	This is the lead of the project the issue belongs to.
Component Lead	This is the lead of the component the issue belongs to.
Single User	This states that any user that exists in Jira.
Group	This states that all users belong to the specified group.
Project Role	This states that all users belong to the specified project role.
Single Email Address	This states any email address.
All Watchers	This states that all users are watching this issue.
User Custom Field Value	This states the users specified in the user-type custom field. For example, if you have a **User Picker Custom Field** called **Recipient**, the user that's selected in the custom field will receive notifications if he/she has access to the issue.
Group Custom Field Value	This states all users that belong to the group in the group-type custom field. For example, if you have a **Group Picker Custom Field** called **Approvers**, all users from the group (with access to the issue) that are selected in the custom field will receive notifications.

As you can see, the list includes a wide range of options from issue reporters to values contained in custom fields. Basically, anything that can be represented as a user such as **Project Lead**, or contains user values such as **User Custom Field Value**, can be chosen to receive notifications.

If a user belongs to more than one notification for a single event, Jira will make sure that only one email will be sent so that the user does not receive duplicates. For a user to receive notifications, the user must have permission to view the issue. The only exception to this is when using the single email address option (we will discuss security in Chapter 9, *Securing Jira*). If the user does not have permission to view the issue, Jira will not send a notification email.

We will look at how you can add notifications to events so that users can start receiving emails shortly; however, before that, you need to first take a look at the notification scheme.

The notification scheme

The notification scheme is a reusable entity that links events with notifications. In other words, it contains the associations between events and their respective email recipients:

1. Browse to the Jira administration console.
2. Select the **Issues** tab and then the **Notification Schemes** option. This will bring up the **Notification Schemes** page:

Notification Schemes ⑦

The table below shows the notification schemes currently configured for this server

Name	Projects	Actions
Default Notification Scheme	• Engineering • Project Management • Sales	Notifications Copy Edit Delete
Demo Notification Scheme	• Demonstration Project	Notifications Copy Edit Delete
Publication Notification Scheme	• Publication	Notifications Copy Edit Delete

Add notification scheme

From this screen, you can see a list of all the notification schemes and the projects that are currently using them.

Jira comes with a generic default notification scheme. The default scheme is set up with notifications set for all the system events. This allows you to quickly enable notifications in Jira. The default setup has the following notifications:

- **Current Assignee**
- **Reporter**
- **All Watchers**

You can modify the default notification scheme to add your own notification rules, but as you grow your Jira adoption, it is a better idea to create a new scheme from scratch or copy the default scheme and make your modifications.

Adding a notification scheme

Unlike other schemes such as the workflow scheme, where Jira will create one whenever a new project is created, all new projects will be set to use the **Default Notification Scheme**. So, if you want to create notifications that are specific to your project, you will have to create a new notification scheme. Perform the following steps to create a new notification scheme:

1. Browse to the **Notification Schemes** page
2. Click on the **Add Notification Scheme** button at the bottom
3. Enter a name and description for the new notification scheme
4. Click on the **Add** button to create the notification scheme

When you create a new notification scheme, you create a blank scheme that can be configured later so that you can add your own notification rules to it. It is important that you configure its notification rules before applying the scheme to projects after you create a new notification scheme; otherwise, no notifications will be sent out. We will look at how to configure notification rules later in this chapter.

Deleting a notification scheme

Unlike most other schemes, such as workflow, Jira allows you to delete notification schemes even when they are being used by projects. However, Jira does prompt you with a warning when you attempt to delete a notification scheme that is in use.

Perform the following steps to delete a notification scheme:

1. Browse to the **Notification Schemes** page
2. Click on the **Delete** link for the notification scheme you wish to remove
3. Click on the **Delete** button to remove the notification scheme

Once you have deleted a notification scheme, the projects that were previously using the scheme will have no notification schemes, so you will have to reapply schemes individually. When you delete a notification scheme, you remove all the notifications you set up in the scheme.

Managing a notification scheme

Notification schemes contain notifications that are set on events in Jira. Perform the following steps to configure a notification scheme:

1. Browse to the **Notification Schemes** page.
2. Click on the **Notifications** link for the notification scheme you wish to configure. This will bring you to the **Edit Notifications** page.

This page lists all the existing events in Jira and their corresponding notification recipients. If you configure a new notification scheme, there will be no notifications set for the events.

Adding a notification

There are two ways you can add a new notification. You can add a notification for a specific event or you can add a notification for multiple events. Perform the following steps to add a new notification:

1. Browse to the **Edit Notifications** page for the notification scheme you wish to configure.
2. Click on the **Add Notification** link or the **Add** link for the event you wish to add a notification for. Both actions will bring you to the **Add Notification** page. If you click on the **Add** link, the **Events** selection list will preselect the event for you.
3. Select the events you want to add the notification type to.
4. Select the notification type from the available options.
5. Click on the **Add** button. For example, the following screenshot shows setting up a notification for Jira to send out emails to the project lead when issues are created and updated:

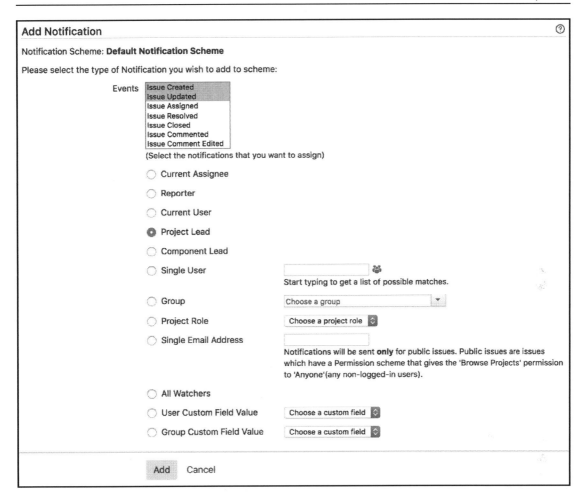

Once added, the notification will be listed against the events that have been selected. You can continue adding notifications for the events by repeating the same steps.

　You can select multiple events to add a notification type to.

Deleting a notification

When notifications are no longer required for certain events, you can also have them removed. To remove notifications, you will need to do so one by one, per event:

1. Browse to the **Edit Notifications** page for the notification scheme you wish to configure
2. Click on the **Delete** link for the notification you wish to remove
3. Click on the **Delete** button to remove the notification for the event

After you remove a notification, users affected by that notification will stop receiving emails from Jira. However, you need to pay attention to your configurations, as there may be other notifications for the same event that will continue to send emails to the same user. For example, if you created two notifications for the issue created event—one set to a single user, John (who belongs to the **jira-administrator** group), and another set to **jira-administrator** group—and your goal is to prevent emails being sent to the user John, you will need to remove both notifications from the event instead of simply using the **Single User** option.

Assigning a notification scheme

When new projects are created, they are automatically assigned to use the default notification scheme. If you want your project to use a different scheme, you will need to go to the **Notifications** section of your project's administration console:

1. Browse to the project administration page for the project you want to apply the notification scheme to.
2. Select the **Notifications** option from the left panel:

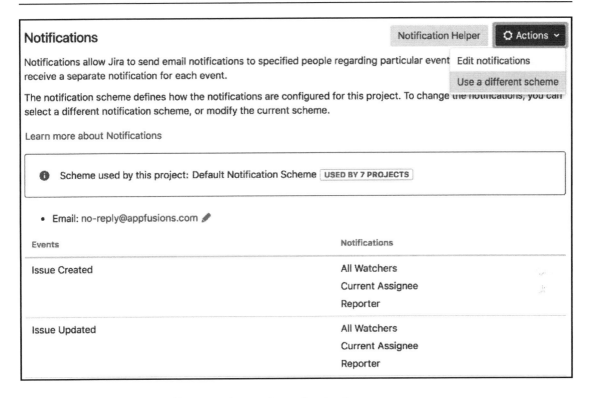

3. Select **Use a different scheme** from the **Actions** menu.

4. Select the notification scheme to use.

5. Click on the **Associate** button.

As soon as a notification scheme is applied to the project, it will take effect immediately, and you will see emails being sent out for the events that have been configured in the scheme. Like any other schemes in Jira, notification schemes can be assigned to multiple projects so that they can share the same notification behavior.

Batching email notifications

One common complaint users have about Jira's email notification is its frequency. Every change made to the same issue will trigger an email to be sent, and for a busy team that's constantly updating issues, such as adding comments, this can very quickly cause a storm of emails being sent that flood user's inboxes. To help alleviate this challenge, Jira 8 has introduced the feature of batch email notifications. The way this works is as follows: all changes that are made to a given issue in the past 10 minutes will be batched into a single summary email, so the user will only receive one email for these changes instead of one per change. This will greatly help reduce the amount of clutter that's generated as a result of frequent updates that are made to issues. To enable email batching, follow these steps:

1. Browse to the Jira administration console
2. Select the **System** tab and then the **Batching email notifications** option
3. Check the **Batching email notifications** option to enable it

As this is a new feature that was added in Jira 8, improvements are being planned for future releases such as adding support to include custom field values to the summary email and having a customizable batching option.

Troubleshooting notifications

Often, when people do not receive notifications from Jira, it can be difficult and frustrating to find the cause. The two most common causes for notification-related problems are either outgoing mail server connectivity or misconfiguration of the notification scheme.

Troubleshooting outgoing mail server problems is quite simple. All you have to do is try to send out a test email, as described in the *Sending a test mail* section. If you receive your test email, then there will be no problems with your outgoing mail server configuration and you can focus on your notification configurations.

Troubleshooting notifications are not as straightforward since there are a number of things that you will need to consider. To help with this challenge, Jira has a handy feature called notification helper. The notification helper can save the Jira administrator time by helping them pinpoint why a given user does or does not receive notifications. All the administrator has to do is tell the helper who the user is, which issue (or an example issue from a project) the user will or will not be receiving notifications for, and the event that is triggering the notification:

1. Browse to the Jira administration console
2. Select the **System** tab and then the **Notification Helper** option

3. Specify the user that will or will not receive notifications in the **User** field
4. Specify the issue to test with
5. Select the type of notification event
6. Click on **Submit**

The **Notification helper** feature will then process the input and report whether the user is expected to receiving notifications, and why, based on notification scheme settings:

As you can see from the preceding screenshot, the user, `Patrick Li`, is currently not receiving notifications for the `DEMO-3` issue when it is being updated because the notification is set up to have only the **Current Assignee** receive emails, and **Patrick Li is not the assignee**.

Incoming emails

We have seen how you can configure Jira to send emails to notify users about updates on their issues. However, this is only half of the story when it comes to Jira's email support.

You can also set up Jira so that it periodically polls mailboxes for emails and creates issues based on the emails' subject and content. This is a very powerful feature with the following benefits:

- It hides the complexity of Jira from business users so that they can log issues more efficiently and leave the complexity to the IT team.
- It allows users to create issues, even if Jira can only be accessed within the internal network. Users can send emails to a dedicated mailbox for Jira to poll.

Adding an incoming mail server

For Jira to retrieve emails and create issues from them, you need to add the POP/IMAP mail server configurations to Jira. POP and IMAP are mail protocols that are used to retrieve emails from the server. Email clients, such as Microsoft Outlook, use one of these protocols to retrieve your emails.

Unlike outgoing mail servers, Jira allows you to add multiple incoming mail servers. This is because while you only need one mail server to send emails, you may have multiple mail servers or multiple mail accounts (on the same server) that people will use to send emails to. For example, you might have one that's dedicated to providing support and another one for sales. It is usually a good idea to create separate mail accounts to make it easier when trying to work out which email can go into which project. For this reason, adding POP/IMAP mail servers can be thought of as adding multiple mail accounts in Jira. Perform the following steps to add an incoming mail server:

1. Browse to the Jira administration console.
2. Select the **System** tab and then the **Incoming Mail** option.
3. Click on the **Add POP/IMAP Mail Server** button.

4. Enter a name and description for the mail server.
5. Select the type of mail service provider. For example, if you are using your own hosted mail service or one of the recognized cloud providers such as Google.
6. Specify the hostname of the POP/IMAP server if you are using your own (custom provider).
7. Enter the username/password credentials for the mail account.
8. Click on the **Add** button to create the POP/IMAP mail server:

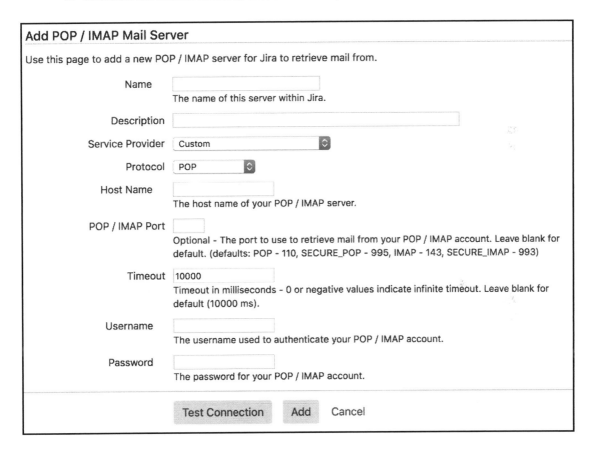

Add POP / IMAP Mail Server

Use this page to add a new POP / IMAP server for Jira to retrieve mail from.

Name

The name of this server within Jira.

Description

Service Provider — Custom

Protocol — POP

Host Name

The host name of your POP / IMAP server.

POP / IMAP Port

Optional - The port to use to retrieve mail from your POP / IMAP account. Leave blank for default. (defaults: POP - 110, SECURE_POP - 995, IMAP - 143, SECURE_IMAP - 993)

Timeout — 10000

Timeout in milliseconds - 0 or negative values indicate infinite timeout. Leave blank for default (10000 ms).

Username

The username used to authenticate your POP / IMAP account.

Password

The password for your POP / IMAP account.

Test Connection   Add   Cancel

You can have multiple incoming servers.

Mail handlers

Mail handlers are what Jira uses to process retrieved emails. Each mail handler is able to process emails from one incoming mail server and periodically scan for new emails.

Jira ships with a number of mail handlers, each with their own features. In the following sections, we will discuss each of the handlers in detail.

Creating a new issue or adding a comment to an existing issue

Creating a new issue or adding a comment to an existing issue mail handler (also known as the create and comment handler in previous versions of Jira) is the most used mail handler. It will create new issues from the received emails and also add comments to existing issues if the incoming email's subject contains a matching issue key. If the subject does not contain a matching issue key, a new issue is created. The following table lists the parameters that are required when creating the mail handler:

Parameter	Description
Project	This is the project in which issues will be created. This is not used for commenting where the email subject will contain the issue key.
Issue Type	This is the issue type for newly created issues.
Strip Quotes	If this is present in the parameters, quoted text from the email will not be added as part of the comment.
Catch Email Address	This specifies whether JIRA is to only handle emails that are sent to the specified address.
Bulk	This specifies how to handle autogenerated emails such as those that are generated by Jira. It is possible to create a loop if JIRA sends emails to the same mailbox where it also picks up emails. To prevent this, you can specify one of the following: **Ignore**: This is used to ignore these emails **Forward**: This is used to forward these emails to another address **Delete**: This is used to delete these emails altogether Generally, you can set it to forward.
Forward Email	If this is specified, then the mail handler is unable to process an email message it receives. An email message indicating this problem will be forwarded to the email address is specified in this field.

Create Users	If the email is sent from an unknown address, Jira will create a new user based on the email *from* address and randomly generate a password. An email will be sent to the *from* address informing the new Jira account user.
Default Reporter	This specifies the username of a default reporter, which will be used if the email address in the **From** field of any of the received messages does not match the address associated with that of an existing Jira user.
Notify Users	Uncheck this option if you do not want Jira to notify new users that are created, as per the **Create Users parameter**.
CC Assignee	Jira will assign the issue to the user that's specified in the **To** field first. If no user can be matched from the **To** field, JIRA will then try the users in the CC and then BCC list.
CC Watchers	JIRA will add users in the CC list (if they exist) as watchers of the issue.

Adding a comment with the entire email body

This mail handler extracts text from an email's content and adds it to the issue with a matching issue key in the subject. The author of the comment is taken from the **From** field.

It has a set of parameters that are similar to the create and comment handler.

Adding a comment from the non-quoted email body

Adding a comment from the non-quoted email body is very similar to the full comment handler, but it only extracts non-quoted text and adds them as comments. Text that starts with > or | is considered to be quoted.

It has a set of parameters that are similar to the create and comment handler.

Creating a new issue from each email message

Creating a new issue from each email message is quite similar to the create and comment handler, except this will always create a new issue for every received email.

It has a set of parameters that are similar to the create and comment handler.

Adding a comment before a specified marker or separator in the email body

Adding a comment before a specified marker or separator in the email body is a more powerful version of the comment handlers. It uses regular expressions to extract text from email contents and adds them to the issue:

Parameter	Description
Split Regex	This regex expression is used to extract contents. There are two rules for the regex expression: Rule 1—it must start and end with a delimiter character, usually with / Rule 2—it cannot contain commas, for example, `/-{}{}{}{}{}\s*Original Message\s*{}-/` or `/_____*/`

Adding a mail handler

You can set up as many mail handlers as you want. It is recommended that you create dedicated mailboxes for each project you wish to allow Jira to create issues from emails. For each account, you will then need to create a mail handler. The mailbox you set up needs to be accessible through POP or IMAP.

Perform the following steps to add a mail handler:

1. Browse to the **Incoming Mail** page.
2. Click on the **Add incoming mail handler** button.
3. Provide a name to the new mail handler.
4. Select an incoming mail server or **Local Files**.
5. Specify how long Jira can wait to poll the mailbox for new emails (in minutes). You will want to keep this long enough to allow enough time for Jira to process all the emails, but not too long as you may end up having to wait for a long time to see your emails converted into issues in Jira.
6. Select the type of handler you want to add.
7. Click on the **Next** button:

Depending on the handler type you select, the next screen will vary. On the next screen, you will need to provide the required parameters for the mail handler, as described in the preceding section. The following screenshot shows an example configuration dialog box, where new issues will be created in the **Help Desk** project as **Task**:

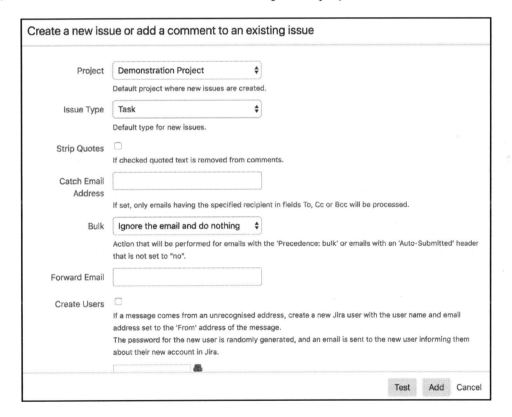

You can always use the **Test** button to test out your configuration. Jira provides helpful hints if there are problems.

Editing and deleting a mail handler

You can update the details of your mail handlers at any time. You will often need to tune your handler parameters a few times until you get your desired results. Perform the following steps to update a mail handler:

1. Browse to the **Incoming Mail** page
2. Click on the **Edit** link for the mail handler you wish to update
3. Update the configure options

Once updated, these changes will be applied immediately and Jira will use the new handler parameters for the next polling run.

You can also delete mail handlers that are no longer required at any time. Perform the following steps to delete a mail handler:

1. Browse to the **Incoming Mail** page
2. Click on the **Delete** link for the mail handler you wish to remove

You will not be prompted with a confirmation page. The mail handler will be removed immediately, so think carefully before you delete it.

Advanced mail handler

The default mail handlers that come with Jira are often enough for simple email processing needs. If you need to have more control or need special processing logic for your incoming emails, you can create custom mail handlers. However, creating new mail handlers requires you to have knowledge of programming; a better option is to use an add-on called **Enterprise Email Handler for Jira (JEMH)**.

With JEMH, you can set up advanced email routing, additional email triggered operations such as updating an issue based on your email content, an audit of received/processed emails, and more. You can find out more about JEMH at `https://marketplace.atlassian.com/apps/4832/enterprise-email-handler-for-jira-jemh`.

The HR project

Users will often want to get progress updates on their issues after they have logged them. So, instead of business users having to ask for updates, we will proactively update them through our newly acquired knowledge, that is, Jira notifications.

In `Chapter 5`, *Field Management*, we added a custom field called `Direct Manager`, which allows users to add the manager of the new employee or leaving employee so that he/she can be kept in the loop.

The other customization we made in `Chapter 7`, *Workflow and Business Process*, was the addition of new transitions in the workflow. We need to make sure that those transitions fire the appropriate events and also send out notifications. In summary, we need to do the following:

- Send out notifications for the events that are fired by our custom workflow transitions
- Send out notifications to users that are specified in our `Direct Manager` custom field

While you can achieve both using other Jira features, such as adding users as watchers to the issue and reusing existing Jira system events, this exercise will explore the options that are available to you. In later chapters, you will see that there are other criteria to consider while deciding on the best approach.

Setting up mail servers

The first step to enable email communication, as you will have guessed, is to register mail servers in Jira. If you are using the standalone distribution of Jira, it is recommended that you add your mail server by entering the host information:

1. Log in to Jira as a Jira administrator.
2. Browse to the Jira administration console.
3. Select the **System** tab and then the **Outgoing Mail** option.

4. Click on the **Configure new SMTP mail server** button.
5. Enter your mail server information. If you do not have a mail server handy, you can sign up for a free Gmail account and use that for testing purposes.

After adding your mail server, you can try sending yourself a quick test email to check whether Jira is able to access your server successfully.

Updating workflow post functions

In `Chapter 7`, *Workflow and Business Process*, we created a few new workflow transitions. We now need to update these new transitions to make sure they fire the appropriate events:

1. Browse to the **View Workflows** page.
2. Click on the **Edit** link for `HR: Termination Workflow`.
3. Click on any transitions other than **Done.**
4. Update the post function to fire the **Issue Updated** event rather than the **Generic Event**.
5. Repeat this for all other transitions except the **Done** transition.
6. Publish the draft workflow. You can save a backup copy in case you want to revert.

We are using the **Issue Updated** event because it reflects the fact that the issue is being updated; also, the event is tied to more appropriate email templates. We can, of course, also create a new custom event and email templates and make the post function fire the custom event instead.

Setting up a notification scheme

Now, you need to have your own notification scheme so that you can start adding notifications to your events. We will base our notification scheme on the default scheme to help us get things set up quickly:

1. Select the **Issues** tab and then the **Notification Schemes** option
2. Click on the **Copy** link for **Default Notification Scheme**
3. Click on the **Edit** link of the copied notification scheme
4. Rename it `HR Notification Scheme` and click on **Update**

This will create a new notification scheme with the basic notifications pre-populated. All you need to do now is modify the events and add your own notification needs.

Setting up notifications

There are two rules you need to follow to add notifications. First, you need to add notifications for your custom events so that emails will be sent out when they are fired. Second, you will want users that are specified in the CC list custom field to also receive emails along with the assignee and reporter of the issue:

1. Click on the **Notifications** link for HR Notification Scheme
2. Click on the **Add notification** link
3. Select the **Issue Updated** event type
4. Select **User Custom Field Value** for the notification type and select Direct Manager from the drop-down list
5. Click on the **Add** button

Nice and easy. With just a few clicks, you have added the Direct Manager custom field to the notification scheme. So now, regardless of who is put into the field, the user will receive notifications for issue updates.

Putting it together

The last step, as always, is to associate your scheme with projects for activation:

1. Browse to the HR project's administration page
2. Select the **Notifications** option from the left panel
3. Select **Use a different scheme** in the **Actions** menu
4. Select the new HR Notification Scheme we just created
5. Click on the **Associate** button

With just a few clicks, you enable Jira to automatically send out emails to update users with their issue's progress. Not only this, but you have tied in the custom fields you created from earlier chapters to manage who, along with the issue assignee and reporter, will also get these notifications. So, let's put this to the test!

1. Create a new **Termination** issue in the HR project.
2. Select a user for the Direct Manager custom field. It is a good idea not to select yourself since the reporter will get notifications by default. Also, make sure that the user that's selected has a valid email address.
3. Transition the issue to move along the workflow.
4. You will receive emails from Jira within minutes.

If you do not receive emails from Jira, check your mail queue and check whether the mail is being generated. Then, follow the steps from the *Troubleshooting notifications* section in this chapter.

Summary

In this chapter, we looked at how Jira can stay in touch with its users through emails. Indeed, with today's new gadgets, such as smartphones and tablets, being able to keep users updated with emails is a powerful feature, and Jira has a very flexible structure in place to define the rules on who will receive notifications.

We also very briefly mentioned some of the security rules about who can receive notifications. Jira performs security checks prior to sending out notifications for two very good reasons—first, there is no point sending out an email to a user who cannot view the issue; second, you will not want unauthorized users viewing the issue and receiving updates that they won't know anything about.

In the next chapter, we will look into the security aspects of Jira and how you can secure your data to prevent unauthorized access.

Section 3: Advanced Jira 8 3

In the final section of this book, you will get exposure to some advanced features of Jira 8. You will explore the Jira security model; understand and learn about the search, report, and analysis functions; and look at Jira Service Desk, which allows Jira to be run as a support portal.

The following chapters will be covered in this section:

- Chapter 9, *Securing Jira*
- Chapter 10, *Searching, Reporting, and Analysis*
- Chapter 11, *Jira Service Desk*

9
Securing Jira

In the previous chapters, you learned how to store data in Jira by creating issues. As you can see, as an information system, Jira is all about data. It should come as no surprise to you that security plays a big role in Jira, to ensure that only the right people will get access to your data. This is achieved by providing a flexible and robust way for you to manage access control. It is able to integrate with your existing security practices, such as a password policy, and allows both centrally controlled permissions and delegating and empowering your project owners and users to manage permissions for their own projects.

By the end of this chapter, you will have learned about the following:

- User directories and how to connect Jira to LDAP
- General access control in Jira
- Managing fine-grained permission settings
- How to troubleshoot permission problems

Before we delve into the deep end of how Jira handles security, let's first take a look at how Jira maintains user and group memberships.

User directories

User directories are what Jira uses to store information about users and groups. A user directory is backed by a user repository system, such as LDAP, a database, or a remote user management system, such as Atlassian Crowd.

You can have multiple user directories in Jira. This allows you to connect your Jira instance to multiple user repositories. For example, you can have an LDAP directory for your internal users and the Jira internal directory using the database for external users. An example is given in the following screenshot, where we have three user directories configured. The first user directory is the built-in Jira **Internal** directory running on the Jira database. The second user directory is connected to the **Microsoft Active Directory (Read Only)** in read-only mode. The last user directory is connected to **Atlassian Crowd**, user identity management software from Atlassian:

User Directories ⑦

The table below shows the user directories currently configured for JIRA.

The order of the directories is the order in which they will be searched for users and groups. Changes to users and groups will be made in the first directory where JIRA has permission to make changes. It is recommended that each user exist only in a single directory.

Directory Name	Type	Order	Operations
Jira Internal Directory You cannot edit this directory because you are logged in through it, please log in as a locally authenticating user to edit it.	Internal		Edit
Active Directory server	Microsoft Active Directory (Read Only)	↑ ↓	Disable | Edit | Test | Synchronize Last synchronized at 1/4/19 7:35 AM (took 10s). Incremental synchronization completed successfully.
Crowd Server	Atlassian Crowd	↑ ↓	Disable | Edit | Test | Synchronize Last synchronized at 1/4/19 7:35 AM (took 5s). Incremental synchronization completed successfully.

Add Directory

Additional Configuration & Troubleshooting

- Directory Configuration Summary

As a Jira administrator, you can manage user directories by performing these two steps:

1. Browse to the Jira administration console
2. Select the **User management** tab and then select the **User Directories** option

From there, you can see the list of user directories you currently have configured in Jira, add new directories, and manually synchronize with the remote user repository.

When adding a new user directory, you need to first decide on the directory type. There are several different user directory types within Jira:

- **Jira internal directory**: This is the built-in default user directory when you first install JIRA. With this directory, all the user and group information is stored in the Jira database.
- **Active directory (AD)/LDAP**: This is used when you want to connect Jira to an LDAP server. With this directory, Jira will use the backend LDAP to query user information and group membership. This is also known as an **LDAP connector** and should not be confused with internal and LDAP authentication directories.
- **Internal with LDAP authentication**: This is also known as a **delegated LDAP**. With this directory type, Jira will only use LDAP for authentication and will keep all user information internally in the database (retrieved from LDAP when the user successfully authenticates for the first time). This approach can provide better performance. Since LDAP is only used for authentication, this avoids the need to download larger numbers of groups from LDAP.
- **Atlassian Crowd**: If you are also using Atlassian Crowd, a user management and **Single Sign-On (SSO)** solution, you can use this directory type to connect to your crowd instance. With this option, you can also configure your Jira instance to participate in the SSO session.
- **Atlassian Jira**: Jira is capable of acting as a user repository for other compatible applications. If you have another Jira instance running, you can use this directory type to connect to the other Jira instance and for user information.

When you have multiple user directories configured for Jira, there are a few important points to keep in mind. The *order* of the user directories is important, as it will directly affect the order Jira will use to search users and apply changes to users and groups. For example, if you have two user directories and both have a user called **admin** with different passwords, this will have the following effects:

- When you log in to Jira with the user admin, you will be logged in as the admin user from the first user directory that is able to validate the password, in the order of listed directories.
- After logging in, you will be granted group membership from the directory that has validated your password. Any other directories will be skipped.
- If you make a change to the admin user, such as changing the full name, then the changes will only be applied to the first directory Jira has write access to.

Another important point to remember when working with user directories is that you cannot make changes to the user directory when you are logged in with a user account that belongs to the said directory. For example, if you are logged in with an LDAP account, then you will not be able to make changes to Jira's LDAP user directory settings, since there is the potential for the new change to actually lock you out of Jira.

> Always have an active administrator user account ready in the default Jira internal directory. For example, the account created during the initial setup. This will provide you with an administrator account that can help you fix user directory problems, such as the preceding scenario. If you have a user account with the same name in the other user directory, then the internal directory should also be the first one in the list.

Connecting to LDAP

Jira supports a wide range of LDAP servers, including Microsoft Active Directory, OpenLDAP, and the Novell eDirectory server. If a particular LDAP is not listed as one of the options, then we also have a **Generic Directory Server** option.

When using the AD/LDAP connector directory type, you can choose to connect with one of the permission options:

- **Read only**: Jira cannot make any modifications to the LDAP server.
- **Read only, with local groups**: Information retrieved from LDAP will be read-only, but you can also add users to groups created within Jira. These changes will not be reflected in LDAP.
- **Read/Write**: Jira will be able to retrieve and make changes to the LDAP server.

The **Read only** option is the most common option, as IT teams often centrally manage LDAP servers and changes are not allowed. With this option, Jira will only need read access to use data stored in LDAP to verify user credentials and group membership. If you only want to use LDAP as a user repository and authentication, but still want to have the flexibility to update group membership without having to get the LDAP team involved, then the **Read only, with local groups** option will be the best fit. Lastly, the **Read/Write** option should be avoided, as propagating changes to LDAP, such as group membership, can have an unforeseen impact on other systems also relying on the same LDAP server.

To connect your Jira to LDAP, all you have to do is add a new user directory as follows:

1. Browse to the **User Directories** page.
2. Click on the **Add Directory** button and select either **Microsoft Active Directory** or **LDAP** from the **Directory Type** select list, and then click on **Next**.
3. Provide your LDAP server information.

Since every LDAP is different, the exact parameters that are required will vary. At a minimum, you need to provide the following information:

Parameter	Description
Name	This is the name of the user directory.
Directory Type	This is where you select the flavor of your LDAP. This will help Jira to prefill some of the parameters for you.
Hostname	This is the hostname of your LDAP server.
Port	This is the port number of your LDAP server. Jira will prefill this based on your directory type selection.
Base DN	This is the root node for Jira to search for users and groups.
LDAP Permissions	This helps you choose whether Jira should be able to make changes to LDAP.
Username	This is the username that Jira will use to connect to LDAP for user and group information.
Password	This is the password that Jira will use to connect to LDAP.

You can see these sections completed in the following screenshot:

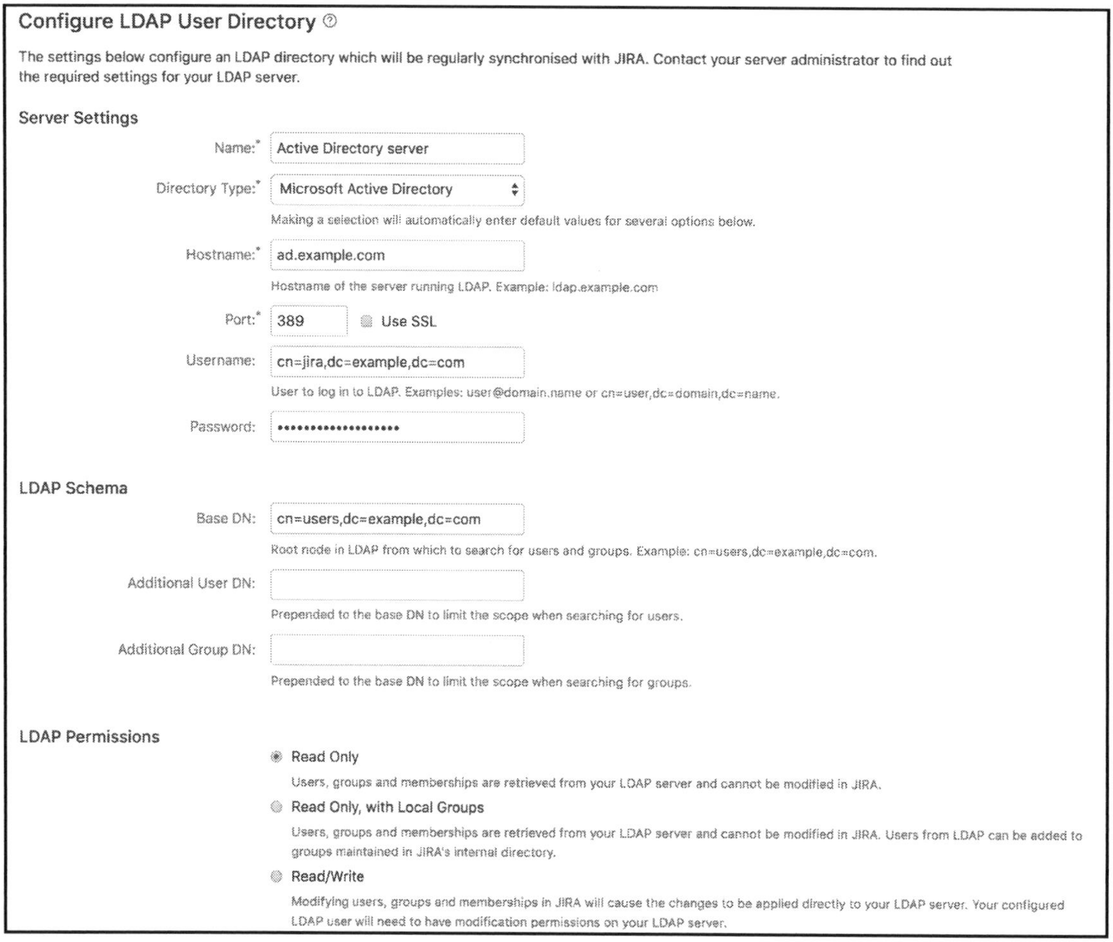

Configure LDAP User Directory ⓘ

The settings below configure an LDAP directory which will be regularly synchronised with JIRA. Contact your server administrator to find out the required settings for your LDAP server.

Server Settings

Name:* `Active Directory server`

Directory Type:* `Microsoft Active Directory ▾`

Making a selection will automatically enter default values for several options below.

Hostname:* `ad.example.com`

Hostname of the server running LDAP. Example: ldap.example.com

Port:* `389` ⚪ Use SSL

Username: `cn=jira,dc=example,dc=com`

User to log in to LDAP. Examples: user@domain.name or cn=user,dc=domain,dc=name.

Password: `••••••••••••••••••`

LDAP Schema

Base DN: `cn=users,dc=example,dc=com`

Root node in LDAP from which to search for users and groups. Example: cn=users,dc=example,dc=com.

Additional User DN:

Prepended to the base DN to limit the scope when searching for users.

Additional Group DN:

Prepended to the base DN to limit the scope when searching for groups.

LDAP Permissions

◉ Read Only

Users, groups and memberships are retrieved from your LDAP server and cannot be modified in JIRA.

◯ Read Only, with Local Groups

Users, groups and memberships are retrieved from your LDAP server and cannot be modified in JIRA. Users from LDAP can be added to groups maintained in JIRA's internal directory.

◯ Read/Write

Modifying users, groups and memberships in JIRA will cause the changes to be applied directly to your LDAP server. Your configured LDAP user will need to have modification permissions on your LDAP server.

Apart from the preceding parameters, there are additional advanced settings, such as **User Schema Settings** and **Group Schema Settings**. After filling in the form, you can click on the **Quick Test** button to verify that Jira is able to connect to your LDAP server and authenticate with the username and password provided. Note that this does not test for things such as the user lookup. If the initial quick test is successful, then you can go ahead and click on the **Save and Test** button. This will add the user directory and take you to the test page, where you can test the settings with a proper user credential (this will be different than the one used by Jira to connect to LDAP):

Test Remote Directory Connection ⑦

Use this form to test the connection to Microsoft Active Directory (Read Only) directory 'Microsoft Directory server'.

For extended testing enter the credentials of a user in the remote directory.

> ⓘ   Test basic connection : Succeeded

> ⓘ   Test retrieve user : Succeeded

> ⓘ   Test get user's memberships : Succeeded, 16 groups retrieved

> ⓘ   Test retrieve group : Succeeded

> ⓘ   Test get group members : Succeeded, 1 users retrieved

> ⓘ   Test user can authenticate : Succeeded

User name	patrick
Password	••••••••••••••

Test Settings   Edit Settings   Back to directory list

After the new user directory is added, Jira will automatically synchronize with the LDAP server and pull in users and groups. Depending on the size of your LDAP server, this may take some time to complete. After the initial synchronization, Jira will periodically perform incremental synchronization for any changes every 60 minutes.

Users

Each user needs to have an account for them to access Jira, unless it is configured to allow anonymous access (by selecting the **Anyone** group in the **Browse Project** permission scheme; refer to the *Permission schemes* section in this chapter for details). Each user is identified by their username.

 In Jira 7, usernames can be changed after the account is created.

User browser

The user browser is where you will be able to see a list of all the users in Jira, including their usernames, email addresses, last login attempt, and which user directory they belong to. The user browser also provides you with search capabilities. You will be able to search for users that fit the criteria, such as username, full name, email address, and group association. Perform the following steps to access the user browser:

1. Browse to the Jira administration console.
2. Select the **User management** tab and then the **Users** option. This will bring up the **User Browser** page.

By default, the results will be paginated to show 20 users per page, but you can change this setting to show up to 100 users per page. When dealing with large deployments having hundreds of users, these options will become very useful to quickly find the users you need to manage.

Other than providing the ability for you to effectively search for users, the user browser also serves as the portal for you to add new users to Jira and manage a user's group/role associations:

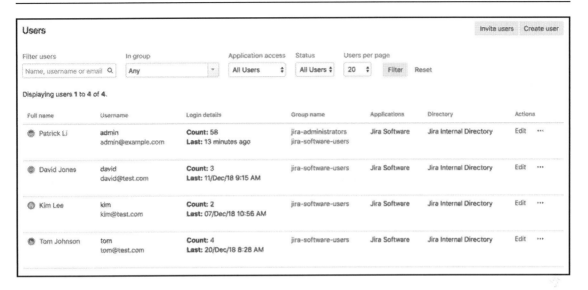

Adding a user

New users can be added to Jira in a number of ways:

- Direct creation by the Jira administrator
- Invitation from the Jira administrator to create an account
- By signing up for an account if the public signup option is enabled
- Synchronization of user accounts from an external user repository, such as LDAP

The first and second options have centralized management, where only the Jira administrators can create and maintain user accounts. This option is applicable to the most private Jira instances designed to be used by an organization's internal users.

The third option allows users to sign up for accounts by themselves. This is most useful when you run a public Jira instance, where manually creating user accounts is not scalable enough to handle the volume. We will be looking at how to enable public signup options in later sections in this chapter. For now, we will examine how administrators can create user accounts manually:

1. Browse to the **User Browser** page.
2. Click on the **Create user** button.
3. Enter a unique username for the new user. Jira will let you know if the username is already taken.

4. Enter the password, full name, and email address of the user. If you do not enter a password, a random password will be generated. In this case, you should select the **Send notification email** option so that the new user can reset their password.

5. Select which applications in Jira the new user will have access to. For example, if you are running Jira software, you should check the **Jira Software** option. Doing so will consume one license seat count.

6. Click on the **Create user** button to create the new user.

Alternatively, the administrator can also choose to invite users so that they can create their accounts themselves. This is different than the public signup option, since only recipients of invitations will be able to create accounts. For this feature to work, you will need to have an outgoing mail server configured, as invitations will be sent as emails. Perform the following steps to invite users to sign up:

1. Browse to the **User Browser** page.
2. Click on the **Invite users** button.
3. Specify the email addresses for the people you wish to invite. You can invite multiple people at once.
4. Click on the **Invite users** button to send out the invitations.

Enabling public signup

If your Jira is public (for example, an open source project), then creating user accounts individually, as explained earlier, would become a very demanding job for your administrator. For this type of Jira setup, you can enable public signup to allow users to create accounts by themselves. Perform the following steps to enable public signup in Jira:

1. Browse to the Jira administration console
2. Select the **System** tab and then the **General configuration** option
3. Click on the **Edit Settings** button
4. Select **Public** for the **Mode** field
5. Click on the **Update** button to apply the setting

Once you have set Jira to run in **Public** mode, users will be able to sign up and create their own accounts from the login page:

As you will see in the *Global permissions* section later in this chapter, once a user signs up for a new account, they will automatically join groups with Jira user's global permission. If you have set Jira to run in **Private** mode, then only the administrator will be able to create new accounts.

Enabling CAPTCHA

If you're running Jira in **Public** mode, you run the risk of having automated spam bots creating user accounts on your system. To counter this, Jira provides the CAPTCHA service, where potential users will be required to type a word represented in an image into a text field. Perform the following steps to enable the CAPTCHA service:

1. Browse to the Jira administration console
2. Select the **System** tab and then the **General configuration** option
3. Click on the **Edit Settings** button
4. Select **On** for the **CAPTCHA on signup** field
5. Click on the **Update** button to apply the setting

Now, when someone tries to sign up for an account, Jira will present them with a CAPTCHA challenge that must be verified before the account is created:

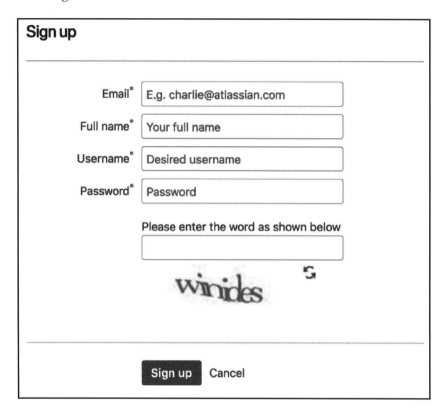

Groups

Groups are a common way of managing users in any information system. A group represents a collection of users, usually based on their positions and responsibilities within the organization. In Jira, groups provide an effective way to apply configuration settings to users, such as permissions and notifications.

Groups are global in Jira, which is something that should not be confused with project roles (which we will discuss later). This means, if you belong to the **jira-administrators** group, then you will always be in that group regardless of which project you are accessing. You will see in later sections how this is different from project roles and their significance.

Group browser

Similar to the user browser, the group browser allows you to search, add, and configure groups within Jira:

1. Browse to the Jira administration console.
2. Select the **User management** tab and then the **Groups** option. This will bring up the **Group Browser** page:

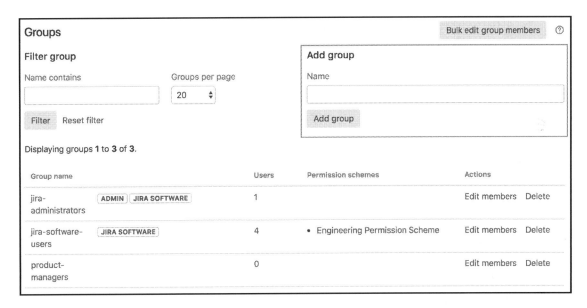

Jira comes with a few default groups. These groups are created automatically when you install Jira. For Jira software, we have the following:

Group	Description
jira-administrators	Administrators of Jira. By default, this group lets you access the administration console.
jira-software-users	By default, members of this group will have access to the Jira Software application.

Other Jira applications, such as Jira Service Desk, will have different sets of default groups.

Adding a group

Other than the three groups that come by default with Jira, you can create your own groups. It is important to note that once you create a group, you cannot change its name. Therefore, make sure that you think about the name of the group carefully before you create it:

1. Browse to the **Group Browser** page
2. Enter a unique name the new group in the **Add group** section
3. Click on the **Add group** button to create the new group

After a group is created, it will be empty and will have no members; you will need to manually add users to the group.

Editing group memberships

Often, people move around within an organization, and your Jira needs to be kept up to date with such movement. In the group browser, there are two ways to manage group memberships. The first option is to manage the membership on a per group level, and the second option is to manage several groups at the same time. These options are actually similar, so we will cover them at the same time.

Perform the following steps to manage individual groups:

1. Browse to the **Group Browser** page.
2. Click on the **Edit members** link for the group you wish to manage members of. This will bring you to the **Bulk Edit Group Members** page.

Perform the following steps to manage multiple groups:

1. Browse to the **Group Browser** page.
2. Click on the **Bulk edit group members** button at the top. This will bring you to the **Bulk Edit Group Members** page.

You will notice that both options will take you to the same page. The difference is that, if you chose the individual group option, Jira will auto select the group to update, and if you chose the bulk edit option, then no groups will be selected. However, regardless of which option you choose, you can still select one or all of the groups to apply your changes to:

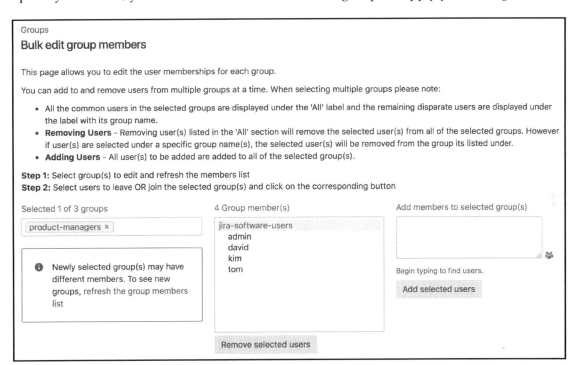

Perform the following steps to update the membership in one or more groups:

1. Browse to the **Bulk Edit Group Members** page
2. Select one or more groups to update
3. Select users from the middle box and click on the **Remove selected users** button to take users out of the groups
4. Specify users (by typing usernames) in the right-hand side box and click on the **Add selected users** button to add users to the groups

Deleting a group

If a group has become redundant, you can remove it from Jira:

1. Browse to the **Group Browser** page
2. Click on the **Delete** link for the group you wish to remove, and click on **Delete** again to permanently remove the group

Once you remove the group, all the users who previously belonged to it will have their group associations updated to reflect the change. However, if you have other configurations using the group, it can have a negative impact if you are not careful. For example, if you are restricting the **Create Issue** project permission to only a group called developers in your permission scheme, by deleting the developers group, nobody will be able to create issues in the projects using the permission scheme.

Be very careful when deleting a group, as it might be used in other configurations.

Project roles

As you have seen, groups are collections of users and are applied globally to all projects in Jira. Jira also offers another way of grouping users, which is applied on the project level only.

Project role browser

Similar to users and groups, project roles are maintained by the Jira administrator through the **Project Role Browser** page. There is a slight difference, however, because since project roles are specific to projects, Jira administrators only define what roles are available in Jira and their default members. Each project's administrators (discussed in later sections) can further define each role's membership for their own projects, overriding the default assignment. We will first look at what Jira administrators can control through the **Project Role Browser** page and then look at how project administrators can fine-tune the membership assignment later.

Perform the following steps to access the **Project Role Browser** page:

1. Browse to the Jira administration console.
2. Select the **System** tab and then the **Project roles** option. This will bring up the **Project Role Browser** page:

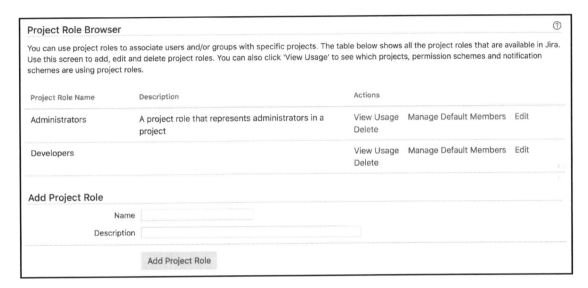

Adding a project role

To start creating your own project roles, you will first need to add the role as a Jira administrator, and then each project administrator will be able to add users to it. Perform the following steps to create a new project role:

1. Browse to the **Project Role Browser** page
2. Enter a unique name for the new project role in the **Add Project Role** section
3. Click on the **Add Project Role** button to create the project role

Once you add a new project role, it will appear for every project.

Managing default members

You can assign default members to project roles, so newly created projects will have project roles assigned to them. Default members are an efficient way for Jira administrators to assign project role members automatically, without having to manually manage each new project as it comes in.

For example, by default, users in the **jira-administrators** group will have the **Administrators** project role. This not only increases the efficiency of the setup by creating a baseline for new projects, but also offers the flexibility to allow modifications to the default setup to cater to unique requirements.

Perform the following steps to set default members for a project role:

1. Browse to the **Project Role Browser** page
2. Click on the **Manage Default Members** link for the project role you wish to edit

The following screenshot shows that the **Developers** project role has a default user (Patrick Li) and a default group (**jira-software-users**):

On this page, you will see all the default members assigned to the selected project role. You can assign default memberships based on individual users or groups.

Perform the following steps to add a default user/group to the project role:

1. Click on the **Edit** link for the default member option (either the user or group).
2. Use the user picker/group picker function to select the users/groups you wish to assign to the project role.
3. Click on the **Add** button to assign the role. The following screenshot shows that the **jira-software-users** group is the default group for the **Developers** project role:

Assign Default Groups to Project Role: Developers

You can add and remove default groups from the project role **Developers** by using the 'Join' and 'Leave' buttons below.

- **<< Return to viewing project role Developers**

Add group(s) to project role:

☐   Groups in Project Role

☐   jira-software-users

[Remove]

[Add]

Once added, any new projects created will have the specified users/groups assigned to the project role. It is important to note that changes to default memberships are only applied to new projects. Existing projects will not retrospectively have the new default members applied.

Assigning project role members

As you have seen, Jira allows you to assign default members to projects when they are created. This might be sufficient for most projects when they start, but changes will often need to be made due to staff movement during the project life cycle. While it is possible for the Jira administrator to continue maintaining each project's membership, it can easily become an overwhelming task, and in most cases, since project roles are specific to each project, it makes sense to delegate this responsibility to the owner of each project.

In Jira, an owner of a project is someone with the **Administer Projects** permission. By default, members of the **Administrators project** role will have this permission. We will see how to manage permissions in Jira in a later section.

As a project administrator, you will be able to assign members to various project roles for your project. You can assign roles from the project administration page, as follows:

1. Browse to the project administration page for the project you want to update.
2. Select the **Users and roles** option from the left-hand panel.
3. Click on the **Add users to a role** link.
4. Start typing the user's username or the group's name. Jira will auto-search for results as you type.

5. Click on the **Add** button once you have found the user/group you want to add:

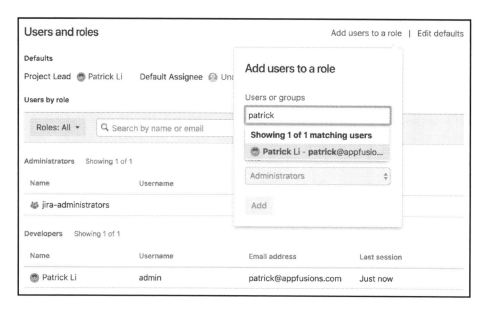

The users and groups assigned to the project role will be for the current project only. Each project administrator can configure this for their own projects. In this way, you can maintain project role memberships separately for each project.

Jira permissions hierarchy

Jira manages its permissions in a hierarchical manner. Each level is more fine-grained than the one above it. For a user to gain access to a resource (for example, to view an issue), they need to satisfy all four levels of permission (if they are all set on the issue in question):

- **Application access**: This defines the groups that will have access to the various applications in Jira (for example, Jira Software)
- **Global permission**: This permission controls access rights functions, such as overall administration
- **Project-level permission**: This permission controls project-level permissions
- **Issue-level security**: This permission controls view access on a per-issue level

We will now look at each of the permission levels and how you can configure them to suit your requirements, starting from the most coarse-grained permission level—global permissions.

Application access

Application access is a new concept introduced in Jira 7, when Jira was divided into Jira Core, Jira Software, and Jira Service Desk. Since each application can have its own license, you, as the administrator, need to have a way of specifying the users who will have access to each application, which translates to the number of license seats that will be consumed. To manage application access, perform the following steps:

1. Browse to the Jira administration console.
2. Select the **Applications** tab and then the **Application access** option.
3. Select the group to grant access to the application. If you check the **Default** option of the group, new users will be added to the group when created:

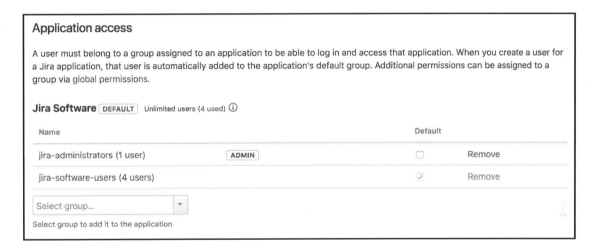

Global permissions

Global permissions, as the name suggests, is the highest permission level in Jira. These are coarse-grained permissions applied globally across Jira, controlling broad security levels, such as the ability to access Jira and administer configurations.

Since they are not fine-grained security, global permissions are applied to groups rather than users. The following table lists all the permissions and what they control in Jira:

Global permission level	Description
Jira System Administrators	This gives permission to perform all Jira administration functions. This is akin to the root mode in other systems.
Jira administrators	This gives permission to perform most Jira administration functions that are not related to system-wide changes. (For example, to configure the SMTP server and to export/restore Jira data.)
Browse Users	This gives permission to view the list of Jira users and groups. This permission is required if the user needs to use the User **Picker/Group Picker** function.
Create Shared Object	This gives permission to share filters and dashboards with other users.
Manage Group Filter Subscriptions	This gives permission to manage group filter subscriptions. Filters will be discussed in `Chapter 10`, *Searching, Reporting, and Analysis*.
Bulk Change	This gives permission to perform bulk operations, including the following: • Bulk edit • Bulk move • Bulk delete • Bulk workflow transition

Jira System Administrator versus Jira Administrator

For people who are new to Jira, it is often confusing when it comes to distinguishing between a Jira System Administrator and a Jira Administrator. For the most part, both are identical, and they can carry out most of the administrative functions in Jira.

The difference is that Jira Administrators cannot access functions that can affect the application environment or network, while a Jira System Administrator has access to everything.

While it can be useful to separate these two, in most cases, it is not necessary. By default, the **jira-administrators** group has both the Jira System Administrators and Jira Administrators permissions.

The following list shows examples of system operations that are only available to people with Jira System Administrators permission:

- Configuring SMTP server details
- Configuring the CVS source code repository
- Configuring listeners

- Configuring services
- Configuring where Jira stores index files
- Importing data into Jira from an XML backup
- Exporting data from Jira to an XML backup
- Configuring attachment settings
- Accessing Jira license details
- Granting/revoking Jira System Administrators global permission
- Deleting users with Jira System Administrators global permission

Configuring global permissions

Global permissions are configured and maintained by Jira Administrators and Jira System Administrators, as follows:

1. Browse to the Jira administration console.
2. Select the **System** tab and then the **Global permissions** option to bring up the **Global Permissions** page, as shown in the following screenshot:

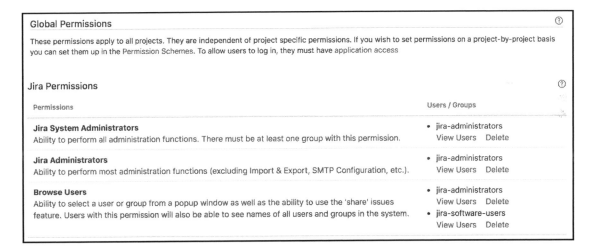

 Users with Jira administrators global permission cannot grant themselves Jira System Administrators global permission.

Granting global permissions

Global permissions can only be granted to groups. For this reason, you will need to organize your users into logical groups for global permissions to take effect. For example, you will want to have all your administrators belong to a single group, such as the **jira-administrators** group, so you can grant them administration permission:

1. Browse to the **Global Permissions** page
2. Select the permission you wish to assign from the **Add Permission** section
3. Choose the group to be given the permission
4. Click on the **Add** button to add the assignment

The **Group** drop-down list will list all the groups in Jira. It will also have an extra option called **Anyone**. This option refers to all users, including those that do not need to log in to access Jira. You cannot select this option when granting the **Jira Users** permission, as they are required to log in, and **Anyone** refers to a non-logged in user. For a production system, it is recommended to take care when granting any global permission to **Anyone** (non-logged in users), as this can lead to security and privacy concerns. For example, by granting **Anyone** as the global permission for **Browse Users**, anyone with access to your Jira instance will be able to get your registered users' information.

Revoking global permissions

Global permissions can also be revoked. Both Jira System Administrators and Jira Administrators can revoke global permissions, but Jira Administrators cannot revoke the Jira System Administrators global permission.

Perform the following steps to delete a global permission from a group:

1. Browse to the **Global Permissions** page
2. Click on the **Delete** link for the group you wish to remove from the global permission from
3. Click on the **Delete** button to revoke the global permission for the group

Jira has built-in validation rules to prevent you from accidentally locking yourself out by mistakenly removing the wrong permissions. For example, Jira will not let you delete the last group from Jira System Administrators global permissions, as doing so effectively prevent you from adding yourself back (since only Jira System Administrators can assign/revoke global permissions).

Project permissions

As you have seen, global permissions are rather coarse in what they control and are applied globally. Since they can only be applied to groups, they are rather inflexible when it comes to deciding whom to grant permissions to.

To provide a more flexible way of managing and designing permissions, Jira allows you to manage permissions on a project level, which allows each project to have its own distinctive permission settings. Furthermore, permissions can be assigned to one of the following:

- **Application access:** This is for any user that has been granted access to the application
- **Reporter**: This is the user who submitted the issue
- **Group**: These are all users that belong to the specified group
- **Single user**: This is any user in Jira
- **Project lead**: This is the lead of the project
- **Current assignee**: This is the user currently assigned to the issue
- **User custom field value**: This user is specified in a custom field of the type **User Custom Field**
- **Project role**: These are all users that belong to the specified role
- **Group custom field value**: These are users within the specified group in a **Group Custom Field**

The list of permissions is also more fine-grained and designed more around controlling permissions on a project level. The only catch to this is that the list is final, and you cannot add new custom permission types:

Project permissions	Description
Administer Project	This is the permission to administer a project. Users with this permission are referred to as project administrators. Users with this permission are able to edit the project role membership, components, versions, and general project details, such as the name and description. Checking the **Extended project administration** option will allow project administrators to make changes to workflows and screens used by their projects.
Browse Project	This is the permission for users to browse and view the project and its issues. If a user does not have the **Browse Project** permission for a given project, the project will be hidden from them, and notifications will not be sent.
Manage Sprints	This is the permission to control who can perform sprint-related operations, such as creating and starting a sprint, on an agile board. This is only applicable to Jira Software.

View Development Tools	This is the permission for users to have access to information from Jira's development tools integration, such as code commits and build results.
View Read-Only Workflow	This is the permission for users to view a read-only diagram of the workflow. When the user has this permission, there will be a **View Workflow** link next to the issue's status.

The issue permissions and their descriptions are as follows:

Issue permissions	Description
Assignable User	This is the user that can be assigned to issues.
Assign Issues	This is the permission for users to assign issues to different users.
Close Issues	This is the permission for users to close issues.
Create Issues	This is the permission for users to create issues.
Delete Issues	This is the permission for users to delete issues.
Edit Issues	This is the permission for users to edit issues.
Link Issues	This is the permission for users to link issues together (if issue linking is enabled).
Modify Reporter	This is the permission for users to change the value of the **Reporter** field.
Move Issues	This is the permission for users to move issues.
Resolve Issues	This is the permission for users to resolve issues and set values for the **Fix For Version** field.
Schedule Issues	This is the permission for users to set and update due dates for issues.
Set Issue Security	This is the permission for users to set issue security levels to enable issue-level security. Refer to upcoming sections to learn more about issue security.
Transition Issues	This is the permission to transition issues through the workflow.

The voters and watchers permissions and their descriptions are as follows:

Voters and Watchers permissions	Description
Manage Watchers	This is the permission to manage the watchers list of issues (add/remove watchers).
View Voters and Watchers	This is the permission to view the voters and watchers list of issues.

The comment permissions and their descriptions are as follows:

Comments permissions	Description
Add Comments	This is the permission for users to add comments to issues.
Delete All Comments	This is the permission to delete all comments.
Delete Own Comments	This is the permission to delete your own comments.
Edit All Comments	This is the permission for users to edit comments made by all users.
Edit Own Comments	This is the permission to edit your own comments.

The attachment permissions and their descriptions are as follows:

Attachments permissions	Description
Create Attachments	This is the permission to add attachments to issues (if an attachment is enabled).
Delete All Attachments	This is the permission to delete all attachments to issues.
Delete Own Attachments	This is the permission to delete attachments to issues added by you.

The time tracking permissions and their descriptions are as follows:

Time tracking permissions	Description
Delete Own Worklogs	This is the permission to delete worklogs made by you.
Delete All Worklogs	This is the permission to delete all worklogs.
Edit Own Worklogs	This is the permission to edit worklogs made by you.
Edit All Worklogs	This is the permission to edit all worklogs.
Work On Issues	This is the permission to log work done on issues (if time tracking is enabled).

As you can see, even though the list cannot be modified, Jira provides you with a very comprehensive list of permissions that will cover almost all your permission needs.

With this many permissions, it would be highly inefficient if you had to create them individually for each project you have. With permission schemes, Jira lets you define your permissions once and apply them to multiple projects.

Permission schemes

Permission schemes, like other schemes, such as notification schemes, are collections of associations between permissions and users or a collection of users. Each permission scheme is a reusable, self-contained entity that can be applied to one or more projects.

Like most schemes, permission schemes are applied on the project level. This allows you to apply fine-grained permissions for each project. Just like project roles, Jira administrators oversee the creation and configuration of permission schemes, and it is up to each project's administrator to choose and decide which permission scheme to use. This way, administrators are encouraged to design their permissions so that they can be reused based on the common needs of their organization. With meaningful scheme names and descriptions, project administrators will be able to choose the scheme that will fit their needs the best instead of requesting a new set of permissions to be set up for each project.

There is, however, some level of freedom for project administrators when it comes to configuration workflows and screens used by their own projects, as we have already seen in earlier chapters covering those two topics. If the permission scheme has the **Extended project administration** option enabled, then any projects using that permission scheme will allow project administrators to make changes without having to rely on a Jira administrator.

We will first look at how Jira administrators manage and configure permission schemes and then how project administrators can apply them in their projects.

Perform the following steps to start managing permission schemes:

1. Browse to the Jira administration console.
2. Select the **Issues** tab and then the **Permission schemes** option to bring up the **Permission schemes** page:

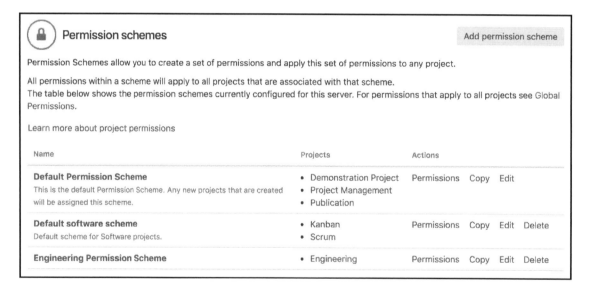

On the **Permission schemes** page, you will see a list of all the permission schemes. From there, you will be able to create new schemes, edit, and delete existing schemes, as well as being able to configure each scheme's permission settings.

Adding a permission scheme

Unlike other schemes, such as workflow schemes, Jira does not create a project-specific permission scheme when you create a new project, but rather, it users a preconfigured scheme called the **Default Permission Scheme**. This scheme is suitable for most simple software development projects. However, it is often not enough, and it is usually a good practice to not modify the **Default Permission Scheme** directly, so you need to create your own permission schemes:

1. Browse to the **Permission Schemes** page
2. Click on the **Add permission scheme** button
3. Enter a name and description for the new permission scheme
4. Click on the **Add** button to create the permission scheme

For new permission schemes, all of the permissions will have no permission configured. This means that if you start using your new scheme straight away, you will end up with a project that nobody can access. We will look at how to configure permissions in later sections of this chapter.

 It is often quicker to clone from an existing permission scheme than to start from scratch.

Configuring a permission scheme

Just like most other schemes in Jira, you need to further fine-tune your permission scheme to make it useful:

1. Browse to the **Permission Schemes** page.
2. Click on the **Permissions** link for the permissions scheme you wish to configure. This will take you to the **Edit Permissions** page.

On this page, you will be presented with a list of project-level permissions, along with short descriptions for each, and the users, groups, and roles that are linked to each of the permissions. You will notice that for the **Default Permission Scheme**, most of the permission options have default users linked to them through project roles. If you are looking at a new permission scheme, there will be no users linked to any of the permissions. This is your one-page view of permission settings for projects, and you will also be able to add and delete users:

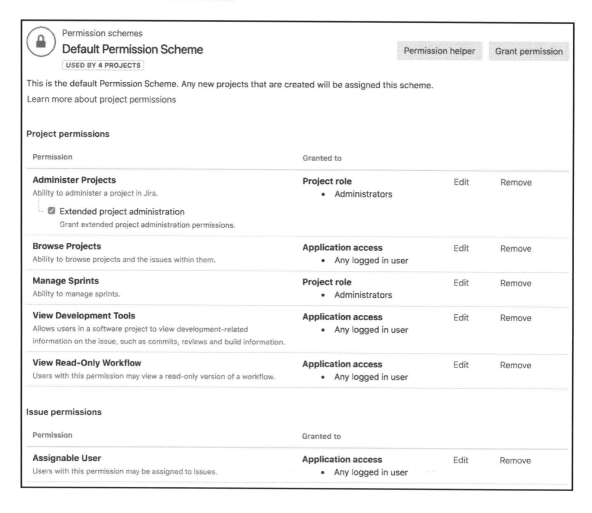

Granting a permissions

As with notification schemes, Jira offers you a range of options for specifying which users should have certain permissions. You can specify users through some of the most common options, such as groups, but you also have some advanced options, such as using users specified in a custom field.

Again, you have two options to grant permissions to a user. You can add them to specific permissions or multiple permissions at once. Both options will present you with the same interface and there is no difference between the two:

1. Browse to the **Edit Permissions** page for the permission scheme you wish to configure.
2. Click on the **Grant permission** button or the **Edit** link for a specific permission.
3. Select the permissions you wish to grant to the user.
4. Select the **User** option to specify whom to grant the permission to. Click on the **Show more** link to see more options.
5. Click on the **Grant** button to grant the selected permission.

Permission options, such as **User Custom Field Value**, are a very flexible way to allow end users to control access. For example, you could have a custom field called **Editors**, and set up your **Edit Issues** permission to allow only users specified in the custom field to be able to edit issues.

The custom field does not have to be placed on the usual view/edit screens for the permission to be applied. For example, you can have the custom field appear on a workflow transition called `Submit to Manager`; once the user has selected the manager, only the manager will have permission to edit the issue.

Revoking permissions

You can easily revoke permissions given to a user, as follows:

1. Browse to the **Edit Permissions** page for the permission scheme you wish to configure
2. Click on the **Remove** link for the permission you wish to revoke, and click on **Remove** again

When you are trying to revoke permissions to prevent users from gaining access to certain things, you need to make sure no other user options are granted the same permission that might be applied to the same user. For example, if you have both the **Single User** and **Group** options set for the **Browse Projects** permission, then you will need to make sure that you revoke the **Single User** option and also make sure that the user does not belong to the **Group** option selected, so you do not have a loophole in your security settings.

Applying a permission scheme

All this time, we have been saying how permission schemes can be selected by project managers to set permissions for their projects; now, we will look at how to apply schemes to your projects. There really is nothing special involved here; permission schemes are applied to projects in the same way as notification and workflow schemes:

1. Browse to the project administration page you want to apply the permission scheme to
2. Select the **Permissions** option from the left-hand panel
3. Select the **Use a different scheme** option from the **Actions** menu
4. Select the **permission scheme** you want to use
5. Click on the **Associate** button

Permission schemes are applied immediately, and you will be able to see the permissions take effect.

Issue security

We have seen how Jira administrators can restrict general access to Jira with global permissions, and what project administrators can do to place fine-grained permissions on individual projects through permission schemes. Jira allows you to take things to yet another level to allow ordinary users to set the security level on the issues they are working with, with issue security.

Issue security allows users to set view permissions (but not edit them) on issues by selecting one of the preconfigured issue security levels. This is a very powerful feature, as it allows the delegation of security control to the end users and empowers them (to a limited degree) to decide who can view their issues.

On a high level, issue security works in a similar way to permission schemes. The Jira administrator will start by creating and configuring a set of issue security schemes with security levels set. Project administrators can then apply one of these schemes to their projects, which allows users (with the **Set Issue Security** project permission) to select the security levels within the scheme and apply them to individual issues.

Issue security schemes

As explained earlier, the starting point of using issue security is issue security schemes. It is the responsibility of the Jira administrator to create and design security levels so that they can be reused as much as possible:

1. Browse to the Jira administration console.
2. Select the **Issues** tab and then the **Issue security schemes** option to bring up the **Issue Security Schemes** page:

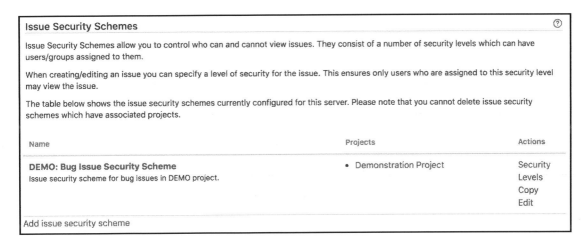

[311]

Adding an issue security scheme

Jira does not come with any predefined issue security schemes, so you will have to create your own from scratch. Perform the following steps to create a new issue security scheme:

1. Browse to the **Issue Security Schemes** page
2. Click on the **Add Issue Security Scheme** button
3. Enter a name and description for the new scheme
4. Click on the **Add** button to create the new issue security scheme

Since an issue security scheme does not define a set of security levels like a permission scheme, you will need to create your own set of security levels right after you create your scheme.

Configuring an issue security scheme

Unlike permission schemes, which have a list of predefined permissions, with issue security schemes, you are in full control of how many options you want to add to schemes.

The options within an issue security scheme are known as **security levels**. These represent the levels of security that users need to meet before Jira will allow them access to the requested issue. Note that, even though they are called security levels, it does not mean that there are any forms of hierarchy among the set of levels you create.

Perform the following steps to configure an issue security scheme:

1. Browse to the **Issue Security Schemes** page.
2. Click on the **Security Levels** link for the issue security scheme you wish to configure. This will bring up the **Edit Issue Security Levels** page:

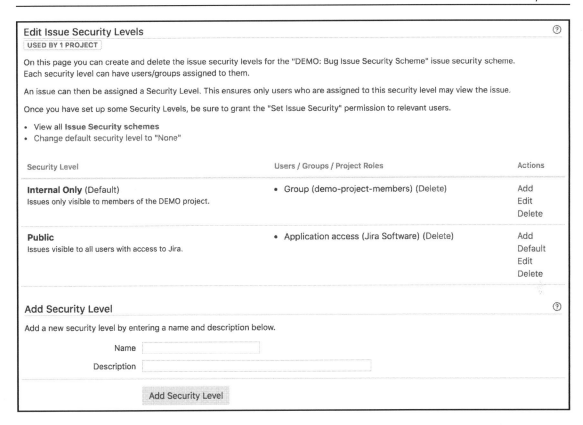

From here, you can create new security levels and assign users to existing security levels.

Adding a security level

Since issue security schemes do not define any security levels, the first step to configure your scheme would be to create a set of new security levels:

1. Browse to the **Edit Issue Security Levels** page for the issue security scheme you wish to configure
2. Enter a name and description for the new security level in the **Add Security Level** section
3. Click on the **Add Security Level** button

You can add as many security levels as you like in a scheme. One good practice is to design and name your security levels based on your team or project roles (for example, developers only).

Assigning users to a security level

Similar to permission schemes, once you have your security levels in place, you will need to assign users to each of the levels. Users assigned to the security level will have permissions to view issues with the specified security level:

1. Browse to the **Edit Issue Security Levels** page
2. Click on the **Add** link for the security level you wish to assign users to
3. Select the option you wish to assign to the security level
4. Click on the **Add** button to assign users

While it may be tempting to use the **Single User** option to add individual users, it is a better practice to use other options, such as **Project Role and Group,** as it is more flexible and doesn't tie the permission to individual users, allowing you to control permission with options such as group association.

Setting a default security level

You can set a security level to be the default option for issues if none are selected. This can be a useful feature for projects with a high-security requirement to prevent users (with the **Set Issue Security** permission) from forgetting to assign a security level for their issues:

1. Browse to the **Edit Issue Security Levels** page
2. Click on the **Default** link for the security level you want to set as default

Once set as default, the security level will have **Default** next to its name. Now, when the user creates an issue and does not assign a security level, the default security level will be applied.

Applying an issue security scheme

Just like permission schemes, project administrators apply issue security schemes to projects. Applying an issue security scheme is similar to applying a workflow scheme, where there is an intermediate migration step involved. This is to ensure that existing issues with set issue security levels can be successfully migrated over to the new security levels in the scheme:

1. Browse to the project administration page you want to apply the workflow scheme to
2. Select the **Issue Security** option from the left-hand panel
3. Select the **Use a different scheme** option form the **Actions** menu
4. Select the permission scheme to use
5. Click on the **Next** button to move to step 2 of the process
6. Select the new security level to apply to the existing issue that may be affected by this change
7. Click on the **Associate** button to apply the new issue security scheme

Troubleshooting permissions

Just like notifications, it can be very frustrating to troubleshoot permission settings. To help with this, Jira also provides a **Permission Helper** to assist administrators with pinpointing settings that prevent users from accessing certain features.

The **Permission Helper** works similarly to the **Notification Helper**:

1. Browse to the Jira administration console
2. Select the **System** tab and then the **Permission helper** option at the bottom
3. Specify the user that is having access problems in the **User** field
4. Specify the issue to test with
5. Select the permission the user does not have (for example, **Edit issue**)

6. Click on **Submit:**

As shown in the preceding screenshot, the user, Patrick Li, cannot edit
issue **DEMO-7** because he does not have the required **Internal Only** issue security level
defined by the issue security scheme used by the project.

Workflow security

The security features we have looked at so far are not applied to workflows. When securing your Jira, you will also need to consider who will be allowed to perform certain workflow transitions. For example, only users in the managers group will be able to execute the authorize transition on issues. For you to enforce security on workflows, you will have to set it on each transition you have by adding workflow conditions. Refer to Chapter 7, *Workflow and Business Process*, which discusses workflows and conditions in more detail.

Password policy

In most cases, unless you are using a single sign-on solution, you will be using a username and password combination to log into Jira. So, you will want users to choose strong passwords that cannot be easily guessed. If your organization is already enforcing a password policy and centrally managing authentication, such as via LDAP, then all you have to do is integrate Jira with it and you are good to go. However, if this is not the case, Jira comes with the ability for you to set a password policy to make sure your users do not choose simple, guessable passwords. To set up a password policy for Jira, perform the following steps:

1. Browse to the Jira administration console
2. Select the **System** tab and then **Password Policy**
3. Choose from one of the available options, with the **Custom** option allowing you to specify your own requirements
4. Click the **Update** button to apply the policy

After you have applied your new password policy, existing user passwords will not be affected. However, if users try to change their own password, or if you are creating a new user account, Jira will make sure the password meets the requirements of the policy, and will display an error message if the requirements are not met:

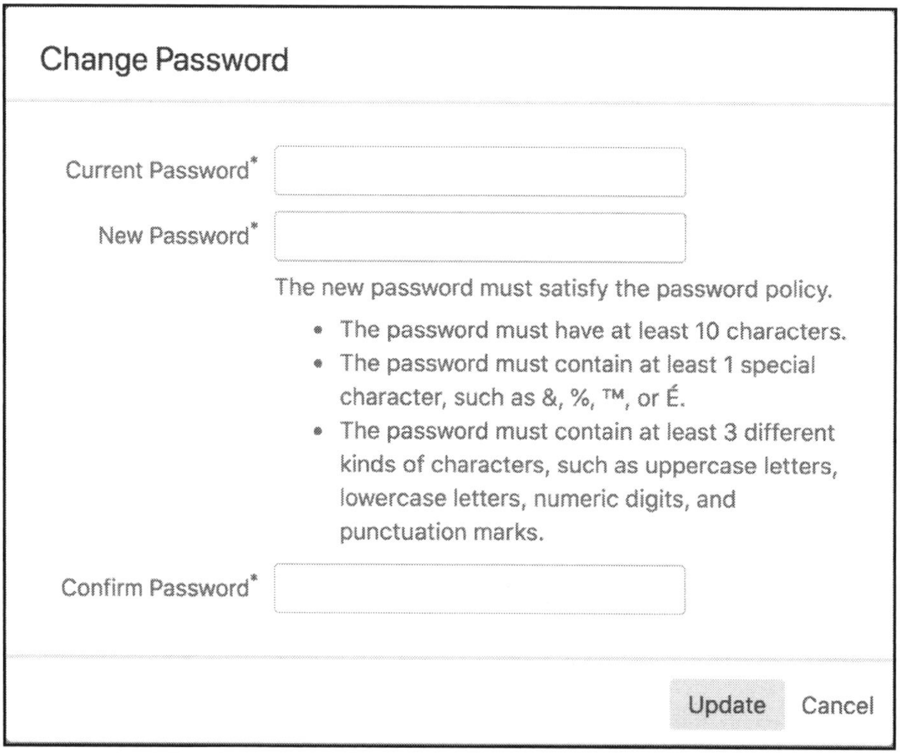

Change Password

Current Password*

New Password*

The new password must satisfy the password policy.

- The password must have at least 10 characters.
- The password must contain at least 1 special character, such as &, %, ™, or É.
- The password must contain at least 3 different kinds of characters, such as uppercase letters, lowercase letters, numeric digits, and punctuation marks.

Confirm Password*

Update   Cancel

Whitelists

In today's world, systems often need to talk to each other to provide a holistic digital experience. While Jira is great at tracking and managing tasks, it is only one of the many systems your users will use on a daily basis to complete their work, so it is important that Jira is able to connect to other systems. If you are running a software engineering team, a good example of this would be integrating Jira with your source control system, such as Atlassian Bitbucket or GitHub, so you can see clearly the branches and code commits that are involved in fixing a bug tracked in Jira.

Of course, you do not want any random systems to connect to Jira, and this is where whitelists comes in. As a Jira administrator, you can specify a list of trusted systems for Jira to connect to, allow those systems to connect to Jira, or both. To whitelist a system, perform the following steps:

1. Browse to the Jira administration console.
2. Select the **System** tab and then **Whitelist.**
3. Enter the URL for the remote system and select how you want Jira to match the URL. You can use regular expressions with the URL.
4. Click the **Add** button to add the URL to the whitelist.

Once the URL is whitelisted, Jira will allow outgoing and incoming (if allowed) connections to the target system.

The HR project

In the previous chapters, you configured your Jira to capture data with customized screens and fields, and processed the captured data through workflows. What you need to do now is secure the data you have gathered to make sure that only authorized users can access and manipulate issues.

Since your HR project is used by the internal team, what you really need to do is put enough permissions around your issues to ensure that the data they hold does not get modified by other users by mistake. This allows us to mitigate human errors by handling access accordingly.

To achieve this, you need to have the following requirements:

- You should know who belongs to the HR team
- Restrict issue assign operations to only the user that has submitted the ticket and members of the HR team
- Do not allow issues to be moved to other projects
- Limit the assignment of tickets to the reporter and members of the HR team

Of course, there are a lot of other permissions we can apply here; the preceding four requirements will be a good starting point for us to build on further.

Setting up groups

The first thing you need to do is to set up a new group for your help desk's team members. This will help you distinguish normal Jira users from your help desk staff:

1. Browse to the **Group Browser** page
2. Name the new group `hr-team` in the **Add Group** section
3. Click on the **Add group** button

You can create more groups for other teams and departments for your scenario here. Since anyone can log a ticket in your project, there is no need to make that distinction.

Setting up user group association

With your group set up, you can start assigning members of your team to the new group:

1. Browse to the **Group Browser** page.
2. Click on the **Edit members** link for the `hr-team` group.
3. Select users with the user picker or simply type in usernames separated by a comma. This time, let's add our admin user to the group.
4. Click on the **Add selected users** button.

Setting up permission schemes

The next step is to set up permissions for our HR project, so you need to have your own permission scheme. As always, it is more efficient to copy the **Default Permission Scheme** as a baseline and make your modifications on top, since we are only making a few changes here:

1. Browse to the **Permission Schemes** pages
2. Click on the **Copy** link for **Default Permission Scheme**
3. Rename the new permission scheme `HR Permission Scheme`
4. Change the description to `Permission scheme designed for HR team projects`

Now that we have our base permission scheme set up, we can start on the fun part, interpreting requirements and implementing them in Jira.

Setting up permissions

The first thing you need to do when you start setting up permissions is to try and match the existing Jira permissions to your requirements. In our case, we want to do the following:

- Restrict who can assign issues
- Restrict who can be assigned to an issue
- Disable issues from being moved

Looking at the existing list of Jira permissions, you can see that we can match the requirements with the **Assign Issues**, **Assignable Users**, and **Move Issues** permissions, respectively.

Once you work out what permissions you need to modify, the next step is to work out a strategy to specify users that should be given the permissions. Restricting the **Move Issue** options is simple. All you have to do is remove the permission for everyone, thus effectively preventing anyone from moving issues in your project.

The next two requirements are similar, as they are both granted to the reporter (the user that submitted the ticket) and our new `hr-team` group:

1. Browse to the **Permission Schemes** pages
2. Click on the **Permissions** link for **HR Permission Scheme**
3. Click on the **Grant permission** button
4. Select both the **Assign Issues** and **Assignable Users** permissions
5. Select the **Reporter** option
6. Click on the **Add** button
7. Repeat the steps and grant the `hr-team` group both permissions

By selecting both the permissions in one go, you have quickly granted multiple permissions to users. Now, you need to remove all the users granted with the **Move Issues** permission. There should be only one granted at the moment, **Any logged in user**, but if you have more than one, you will need to remove all of them:

1. Browse to the **Permission Schemes** page
2. Click on the **Permissions** link for `HR Permission Scheme`
3. Click on the **Remove** link for all the users that have been granted the **Move Issues** permission

That's it! You've addressed all your permission requirements with just a few clicks.

Putting it together

Last, but not least, you can now put on your project administrator's hat and apply your new permission scheme to your HR project:

1. Browse to the project administration page for your HR project
2. Click on the **Permissions** option and select your new HR Permission Scheme
3. Click on the **Associate** button

By associating the permission scheme with your project, you have applied all your permission changes. Now, if you create a new issue or edit an existing issue, you will notice that the list of assignees will no longer include all the users in Jira.

Summary

In this chapter, we first looked at how we can integrate Jira with user repositories, such as LDAP, through user directories. We then looked at Jira's user management options with groups and project roles. Though both are very similar, groups are global, while project roles are specific to each project. We also covered how Jira hierarchically manages permissions. We discussed each permission level in detail and how to manage them.

In the next chapter, we will take a different approach and start looking at another powerful use for Jira—getting your data out through reporting.

10
Searching, Reporting, and Analysis

From Chapter 2, *Using Jira for Business Projects*, to Chapter 6, *Screen Management*, we have looked at how Jira can be used as an information system to gather data from users. In Chapter 7, *Workflow and Business Process*, and Chapter 8, *Emails and Notifications*, we discussed some of the features that Jira provides to add value to the gathered data through workflows and notifications. In this chapter, we will look at the other half of the equation—getting the data out and presenting it as useful information to the users.

By the end of this chapter, you will have covered the following topics:

- Utilizing the search interface in Jira
- Learning the different search options available in Jira
- Getting to know about filters and how you can share search results with other users
- Generating reports in Jira
- Sharing information with dashboards and gadgets

Search interface and options in Jira

As an information system, Jira comes loaded with features and options to search for data and present the search results to end users. Jira comes with three options to perform searches:

- **Quick/text search**: This allows you to search for issues quickly through simple text-based search queries
- **Basic/simple search**: This lets you specify issue field criteria via intuitive UI controls
- **Advanced search**: This lets you construct powerful search queries with Jira's own search language, **Jira Query Language (JQL)**

However, before we start looking into the details of all the search options, let's first take a look at the main search interface that you will be using in Jira while performing your searches.

Issue navigator

The issue navigator is the primary interface through which you will be performing all of your searches in Jira. You can access it by clicking on the **Issues** menu in the top menu bar and then selecting **Search for issues**.

The issue navigator is divided into several sections. The first section is where you will specify all of your search criteria, such as the project you want to search in and the issue type you are interested in. The second section displays the results of your search. The third section includes the operations that you can perform on the search results, such as exporting them into various formats. The fourth and last section lists a number pre-configured, and user-created filters.

When you access the issue navigator for the first time, you will be in the basic search mode. The following screenshot shows the issue navigator in this mode:

In basic search, as you can see, you specify your search criteria with UI controls, selecting the values for each field.

If you previously visited the issue navigator and chose to use a different search option, such as advanced search, then Jira will remember this and open up advanced search instead.

Basic search

This is also known as simple search. The basic search interface lets you select the fields you want to search with, such as issue status, and specify the values for these fields.

With basic search, Jira will prompt you for the possible search values for the selected field. This is very handy for fields, such as status, and select list-based custom fields, so you do not have to remember all the possible options.

For example, as shown in the following screenshot, we are searching for issues in the **Demonstration Project**, with the status as **Open**. And for the status field, Jira will list all the available statuses:

While working with the basic search interface, Jira will have the default fields of project, issue type, status, and assignee visible. You can add additional fields to the search by clicking on the **More** drop-down option and then selecting the field you want to use in the search. Perform the following steps to construct and run a basic search:

1. Browse to the **Issue Navigator** page. If you do not see the basic search interface, and the **Basic** link is showing, click on it to switch to basic search.
2. Select and fill in the fields in the basic search interface. You can click on **More** to add more fields to the search criteria.

Jira will automatically update the search results every time you make a change to the search criteria.

 When working with basic search, one thing to keep in mind is that the project and issue type, context of the custom fields are taken into consideration. (Please see Chapter 5, *Field Management*, for field configuration). If a custom field is set to be applicable to only specific projects and/or issue types, you have to select the project and issue type as part of your search for the custom field to show up.

Advanced search with JQL

Basic search is useful and will fulfill most of the user's search needs. However, there are still some limitations. One such limitation is that basic search allows you to perform searches based on inclusive logic but not exclusive logic. For example, if you need to search for issues in all but one project, you will have to select every project except for the one to be excluded since the basic search interface does not let you specify exclusions.

This is where advanced search comes in. With advanced search, instead of using a field selection-based interface, you will be using what is known as the **Jira Query Language (JQL)**.

JQL is a custom query language developed by Atlassian. If you are familiar with the **Structured Query Language (SQL)** used by databases such as MySQL, you will notice that JQL has a similar syntax; however, JQL is not the same as SQL.

One of the most notable differences between JQL and SQL is that JQL does not start with a select statement. A JQL query consists of a field followed by an operator, and then by a value such as `assignee = john` or a function (which will return a value) such as `assignee = currentUser()`.

You cannot specify what fields to return from a query with JQL, which is different from SQL. You can think of a JQL query as the part that comes after the `where` keyword in a normal SQL `select` statement. The following table summarizes the components in JQL:

JQL component	Description
Keyword	Keywords in JQL are special reserved words that do the following: • Join queries together, such as AND • Determine the logic of the query, such as NOT • Have special meaning, such as NULL • Provide specific functions, such as ORDERBY
Operator	Operators are symbols or words that can be used to evaluate the value of a field on the left and the values to be checked on the right. Examples include the following: • Equals: = • Greater than: > • IN: When checking whether the field value is in one of the many values specified in parentheses
Field	Fields are Jira system and custom fields. When used in JQL, the value of the field for issues is used to evaluate the query.
Functions	Functions in JQL perform specific calculations or logic and return the results as values that can be used for evaluation with an operator.

Each JQL query is essentially made up of one or more components. A basic JQL query consists of the following three elements:

- **Field**: This can be an issue field (for example, a status) or a custom field.
- **Operator**: This defines the comparison logic (for example, = or >) that must be fulfilled for an issue to be returned in the result.
- **Value**: This is what the field will be compared to; it can be a literal value expressed as text (for example, `Bug`) or a function that will return a value. If the value you are searching for contains spaces, you need to put quotes around it, for example, `issuetype = "New Feature"`.

Queries can then be linked together to form a more complex query with keywords such as logical `AND` or `OR`. For example, a basic query to get all the issues with a status of `Resolved` will look similar to this:

```
status = Resolved
```

A more complex query to get all the issues with a `Resolved` status, a `Bug` issue type, and which are assigned to the currently logged-in user, will look similar to the following (where `currentUser()` is a JQL function):

```
issuetype = Bug and status = Resolved and assignee = currentUser()
```

Discussing each and every JQL function and operator is out of the scope of the book, but you can get a full reference by clicking on the **Syntax Help** link in the advanced search interface. The full JQL syntax reference can be found at `https://confluence.atlassian.com/x/ghGyCg`.

You can access the advanced search interface from the **Issue Navigator** page, as follows:

1. Browse to the **Issue Navigator** page
2. Click on the **Advanced** link on the right
3. Construct your JQL query
4. Click on the **Search** button or press the *Enter* key on your keyboard

As JQL has a complex structure and it takes some time to get familiar with, the advanced search interface has some very useful features to help you construct your query. The interface has an **autocomplete** feature (which can be turned off from the **General Configuration** setting) that can help you pick out keywords, values, and operators to use.

It also validates your query in real time and informs you whether your query is valid, as shown in the following screenshot:

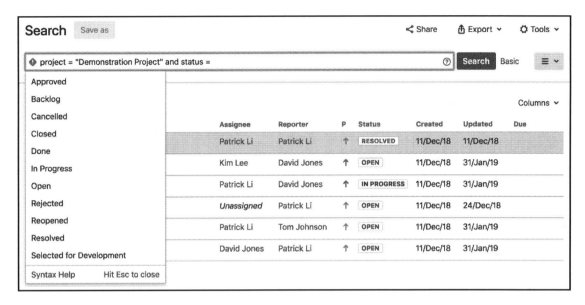

If there are no syntax errors with your JQL query, Jira will display the results in a table below the JQL input box.

You can switch between basic and advanced search by clicking on the **Basic/Advanced** link while running your queries, and Jira will automatically convert your search criteria into and from JQL. In fact, this is a rather useful feature and can help you learn the basic syntax of JQL when you are first starting up, by first constructing your search in basic and then seeing what the equivalent JQL is.

> Switching between simple and advanced search can help you get familiar with the basics of JQL.

You need to take note, however, that not all JQLs can be converted into basic search since you can do a lot more with JQL than with the basic search interface.
The **Basic/Advanced** toggle link will be disabled if the current JQL cannot be converted to the basic search interface.

Quick search

Jira also provides a quick search functionality, which allows you to perform quick simple searches based on the text contained in the issue's summary, description, or comments. This allows you to perform quick text-based searches on all issues in Jira.

The quick search function has several additional features to let you perform more specialized searches with minimal typing, through smart querying. Jira has a list of built-in queries, which you can use as your quick search terms to pull up issues with a specific issue type and/or status. Some useful queries are included in the following table (you can find the full quick search reference at `https://confluence.atlassian.com/jiracoreserver/quick-searching-939937704.html`):

Smart query	Result
Issue key (for example, `HD-12`)	Takes you directly to the issue with the specified issue key.
Project key (for example, `HD`)	Displays all the issues in the project specified by the key in the **Issue Navigator** page.
`my` or `my open bugs`	Displays all the issues that are assigned to the currently logged-in user.
`overdue`	Displays all the issues that are due before today.
Issues with a particular status (for example, `open`)	Displays all issues with the specified status.
Issues with a particular resolution (for example, `resolved`)	Displays all issues with the specified resolution.

You can combine these queries together to create quick yet powerful searches in Jira. For example, the following query brings back all the resolved issues in the `Help Desk` project, where `HD` is the project key:

```
HD resolved
```

Running a quick search is much simpler than either basic or advanced searches. All you have to do is type in either the text you want to search with or the smart query in the **Quick Search** box in the top-right corner, and press *Enter* on your keyboard.

As you can see, the goal of quick search is to allow you to find what you are looking for in the quickest possible way. With smart query, you are able to perform more than just simple text-based searches.

It is important to note that quick search is case sensitive. For example, searching with My instead of my will become a simple text search rather than issues that are assigned to the currently logged-in user.

Working with search results

Now that we have seen how to perform searches in Jira, we will look at different ways we can use and work with the search results, starting with the various features and operations available directly from the issue navigator.

The issue navigator is capable of more than letting you run searches and presenting you with the results. It also has other features which allow you to do the following:

- Display search results in different view options
- Export search results in different formats
- Select the columns you want to see for the issues in the results
- Share your search results with other people
- Create and manage filters

Switching result views

The issue navigator can display your search results in two different views. The default view is **Detail View**, where issues from results are listed on the left-hand side, and the currently selected issue's details are displayed on the right. This view allows you to select and view detailed content of an issue, as well as edit the issue.

The second view is **List View**, where issues are listed in a tabular format. The issue's field values are displayed as table columns. As you will see later, you can configure the table columns as well as the way they are ordered. You can switch between the two views by selecting the options from the **Views** menu next to the **Basic/Advanced** option.

Exporting search results

From the issue navigator, Jira allows you to export your search results to a variety of formats, such as MS Word and CSV. Jira is also able to present your search results in different formats, such as XML or print-friendly pages.

When you select formats such as Word or Excel, Jira will generate the appropriate file and let you download it directly. Perform the following steps to export your results to a different format:

1. Browse to the **Issue Navigator** page
2. Execute a search
3. Select the **Export** drop-down menu in the top-right corner
4. Select the format you wish to see your search results in

Depending on the format you select, some formats will be on screen (example, printable), while others will prompt you with a download dialog box (example, Excel).

Customizing the column layout

If you are using the **List View** option to display your search results, you can configure the field columns to be displayed. In Jira, you can customize your issue navigator for all your personal searches and also on a per-search level with filters (which we will discuss later in this chapter). If you are an administrator, you can set a default column layout for all users (which can be overwritten by each user's own column layout settings).

Perform the following steps to customize your global issue navigator's column layout:

1. Browse to the **Issue Navigator** page.
2. Change your result view to **List View**.
3. Select the **Columns** drop-down menu and a column layout option, as shown here:

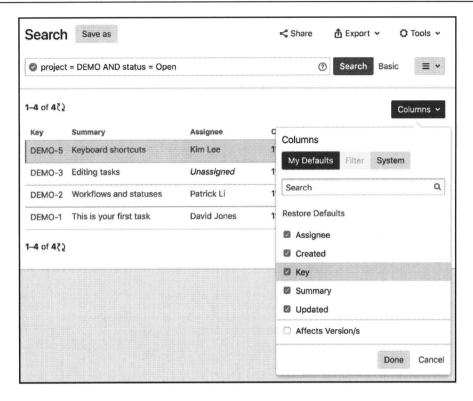

The following options can be used to lay out the columns:

- **My Defaults**: This column layout will be applied to all your searches
- **Filter**: This column layout will be applied only to the current filter
- **System**: This column layout will be applied to all searches

To add or remove a field column, simply check or uncheck the field from the list. To reorder the column layout, you can drag the columns left or right to their appropriate locations.

Sharing search results

After completing a search, you may want to share the results with your colleagues. Now you can tell your colleagues to run the same search or, as we will see later in the chapter, save your search as a filter and then share it with other people. Alternatively, a more convenient way is to use the built-in share feature, especially if this is a one-off share.

To share your current search results, all you have to do is click on the **Share** button in the top-right corner and type in the user's name or an email address (if they are not Jira users). You can add multiple users or email addresses, so you can share this with more than one person. You can also add a quick note, letting people know why you are sharing the search results with them, and Jira will send out emails to all the selected users and email addresses.

Filters

After you have run a search query, sometimes it is useful to save the query for later use. For example, you may have created queries for several projects, listing all the open bugs and new features in each project that are to be completed by a certain date, so you can keep an eye on their progress.

Instead of recreating this search query every time you want to check up on the statuses, you can save the query as a filter, which can be reused at a later stage. You can think of filters as named search queries that can be reused.

Other than being able to quickly pull up a report without having to recreate the queries, saving search queries as filters provides you with other benefits:

- Sharing saved filters with other users
- Using the filters as a source of data to generate reports
- Using the filters for agile boards (see `Chapter 3`, *Using Jira for Agile Projects*)
- Displaying results on a dashboard as a gadget
- Subscribing to the search queries to have results emailed to you automatically

A few things to keep in mind when creating and using filters as the data source for gadgets and agile boards are as follows:

- When you are creating a filter for agile boards, make sure you select the relevant projects as part of your search query.
- When you are creating a filter for gadgets and agile boards, make sure you share the filter with the same group of users that has access to the gadgets and boards. Otherwise, they will not see any results.

We will explore all of the advanced operations you can perform with filters, and explain some of the new terms and concepts, such as dashboard and gadgets, in later sections. However, let's look at how we can create and manage filters first.

Creating a filter

To create a new filter, you will first have to construct and execute your search query. You can do this with any of the three available search options provided in Jira, but please note that the search result must bring you to the **Issue Navigator** page. If you are using the quick search option and search by issue key, you will not be able to create a filter.

Once you have executed your query, regardless of whether it brings back any result, you will be able to create a new filter based on the executed search:

1. Browse to the **Issue Navigator** page
2. Construct and execute a search query in Jira
3. Click on the **Save as** button at the top
4. Enter a meaningful name for the filter
5. Click on the **Submit** button to create the filter

Once you have created the filter, all your search parameters will be saved. In the future, when you rerun the saved filter, Jira will retrieve the updated results based on the same parameters.

Take note that you need to click on the **New filter** button to start a new search if you have created a filter. Since the issue navigator remembers your last search, if you were working with an existing filter, without starting a new search, you will, in fact, be modifying the current filter instead.

Managing filters

As the number of created filters grows, you will need a centralized location to manage and maintain them. To access the **Manage Filters** page.

You can access the page through the issue navigator, as follows:

1. Browse to the **Issue Navigator** page.
2. Click on the **Find filters** link at the left-hand side. You can also access the **Manage Filters** page by going through the top navigator bar.
3. Bring up the drop-down menu from **Issues.**
4. Click on the **Manage filters** option at the bottom of the list.

The **Manage Filters** page displays the filters that are visible to you in three main categories, as set out in the tabs to the left, along with the option to search for existing filters:

Manage Filters			

My Filters ⑦

Filters are issue searches that have been saved for re-use. This page shows all filters that you own.

Name	Shared With	Subscriptions	
☆ Approved (ENGINEERING)	• **Project:** Engineering (VIEW)	None - Subscribe	⚙ ˅
☆ Demo Project Filter	• Private filter	None - Subscribe	⚙ ˅
★ Due this week (HD)	• Private filter	None - Subscribe	⚙ ˅
★ Due this week (HR)	• **Project:** Human Resource (VIEW)	None - Subscribe	⚙ ˅
☆ Highest priority and open (PUB)	• Private filter	None - Subscribe	⚙ ˅
☆ Open and unassigned (PM)	• **Project:** Project Management (VIEW)	None - Subscribe	⚙ ˅

*Tabs on left: Favorite, **My**, Popular, Search*

- **Favorite**: This option lists filters with a gray star next to their names. These filters will be listed in the **Issues** drop-down menu. You can mark a filter as a favorite by clicking on the star directly.
- **My**: This option lists the filters that are created by you.
- **Popular**: This option lists the top 20 filters that have the most people marking them as favorite.
- **Search**: This option searches for existing filters that are shared by other users.

The preceding screenshot also shows that both the **Due this week (HD)** and **Due this week (HR)** filters are marked as favorites.

Sharing a filter

After creating a filter, you can update its details such as name and description, sharing permission, and search parameters. By default, newly created filters are not shared, which means they are only visible to you. To share your filters with other users, perform the following steps:

1. Browse to the **Manage Filters** page
2. Click on the **Edit** option for the filter you wish to edit
3. Update the details of the filter

4. Select the group/project role to share the filter with, and click **Add**
5. Click on the **Save** button to apply the changes

This is shown in the following screenshot:

 Make sure you click on the **Add** link after you have chosen a group or a project to share the filter with.

For you to be able to share a filter, you will also need to have the **Create Shared Object** global permissions (please refer to Chapter 9, *Securing Jira*, for more information on global permissions).

When you share your filter, you can choose who will have view permission, which means see and run your filter (search results are still controlled by project and issue permissions), and who will be able to make modifications to your filter. As we will see later, Jira administrators can also change the ownership of filters that are shared.

Subscribing to a filter

You have seen in Chapter 8, *Emails and Notifications*, that Jira is able to send out emails when certain events occur to keep the users updated. With filters, Jira takes this feature one step further by allowing users to subscribe to a filter.

When you subscribe to a filter, Jira will run a search based on the filter and send you the results in an email. You can specify the schedule of when and how often Jira should perform this. For example, you can set up a subscription to have Jira send you the results every morning before you come to work so when you open up your mail inbox, you will have a full list of issues that require your attention.

To subscribe to a filter, you will need to be able to see the filter (either created by you or shared with you by other users):

1. Browse to the **Manage Filters** page.
2. Locate the filter you wish to subscribe to.
3. Click on the **Subscribe** link for the filter.
4. Select the recipient of the subscription. Normally, this will be you (**Personal Subscription**). You can also create subscriptions for other people by selecting a group.
5. Check the **Email this filter even if there are no issues found** option if you wish to have an email sent to you if there are no results returned from the filter. This can be useful to make sure that the reason you are not getting emails is not other errors.
6. Specify the frequency and time when Jira can send you the emails.

This is shown in the following screenshot:

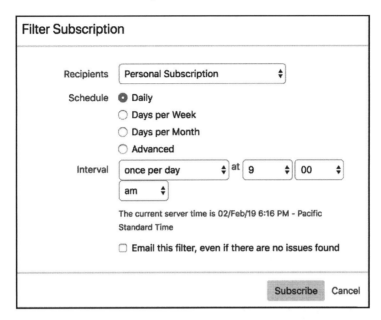

7. Click on the **Subscribe** button. This will create the subscription and take you back to the **Manage Filters** page. The **Subscribe** link will increment the number of subscriptions; for example, **1 Subscription**.

8. Click on the **1 Subscription** link to verify the subscription is created correctly.

9. Click on the **Run now** link to test your new subscription:

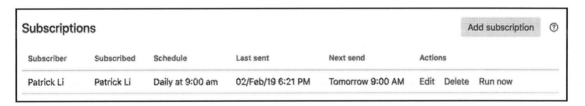

You should see something like the above on your subscription screen once you are done.

Deleting a filter

You can delete a filter when it is no longer needed. However, since you can share your filters out with other users and they can create subscriptions, you need to keep in mind that the filter might be used by an agile board or in other places, as well as the fact that you may impact other users if the filter is shared. Luckily, when you delete a filter, Jira will inform you if other people are using the filter:

1. Browse to the **Manage Filters** page.
2. Click on the **Delete** link for the filter you wish to remove. This will bring up the **Delete Filter** confirmation dialog box.
3. Make sure that the removal of the filter will not impact other users.
4. Click on the **Delete** button to remove the filter.

Jira will inform you if the filter is being used by Jira or there are users subscribed to it. You can click through to see the list of subscribers, and then decide to either proceed with deleting the filter and letting the other users know or leave the filter in Jira.

Changing the ownership of a filter

Usually, Jira only allows the filter's owner to make changes to it unless the right to edit is given to other users. This is usually not a problem for private filters, but when a filter is shared with other users or used for agile boards or dashboard gadgets, this can be problematic when the owner leaves the organization.

For this reason, the Jira administrator is able to change the ownership of a shared filter. Perform the following steps to change a filter's ownership:

1. Browse to the Jira Administration console
2. Select the **System** tab and then the **Shared filters** option
3. Search for the filter you wish to change ownership of
4. Click on the **Change Owner** option
5. Search and select the user that will be the new owner
6. Click on the **Change Owner** button

This is shown in the following screenshot:

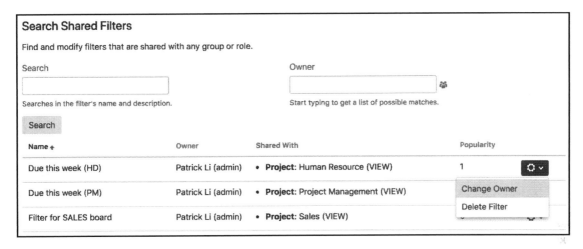

Reports

Apart from JQL and filters, Jira also provides specialized reports to help you get a better understanding of the statistics for your projects, issues, users, and more. Most reports in Jira are designed to report on issues from a specific project; however, there are some reports that can be used globally across multiple projects, with filters.

Generating a report

All Jira reports are accessed from the **Browse Project** page of a specific project, regardless of whether the report is project-specific or global. The difference between the two types of reports is that a global report will let you choose a filter as a source of data, while a project-specific report will have its source of data predetermined based on the project you are in.

When generating a report, you will often need to supply several configuration options. For example, you may have to select a filter, which will provide the data for the report, or select a field to report on. The configuration options vary from report to report, but there will always be hints and suggestions to help you work out what each option is.

Perform the following steps to create a report; you will first need to get to a project's browse page:

1. Select the project you wish to report on.
2. Click on the **Reports** option from the left panel.
3. Select the report you wish to create. The types of reports available will vary depending on the project type.
4. Specify the configuration options for the report.
5. Click on the **Next** button to create the report.

Jira Software comes with a number of reports that are specifically designed around reporting on agile projects, such as the **Burndown Chart**, as well as the basic set of charts that come with Jira Core, such as **Average Age Report** and **Pie Chart Report**. The type of reports available to you depends on the type of project. Scrum and Kanban projects will have reports under the **Agile** category, as shown in the following screenshot:

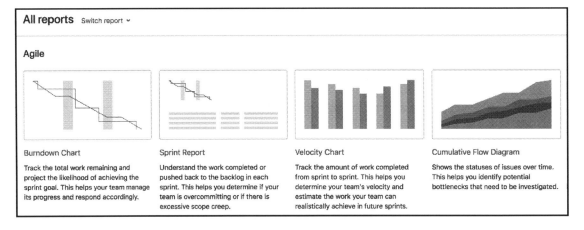

Let's create a pie chart report.

1. We will first select the type of report to be generated by selecting it from a list of available report types that come with Jira, as shown in the following screenshot:

Issue analysis

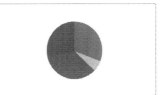

Average Age Report

Shows the average age of unresolved issues for a project or filter. This helps you see whether your backlog is being kept up to date.

Created vs. Resolved Issues Report

Maps created issues versus resolved issues over a period of time. This can help you understand whether your overall backlog is growing or shrinking.

Pie Chart Report

Shows a pie chart of issues for a project/filter grouped by a specified field. This helps you see the breakdown of a set of issues, at a glance.

2. We will then configure the necessary report parameters. In this case, you need to specify whether you are generating a report based on a project or an existing filter; by default, the current project will be preselected. You also need to specify which issue field you will be reporting on, as you can see from the following screenshot:

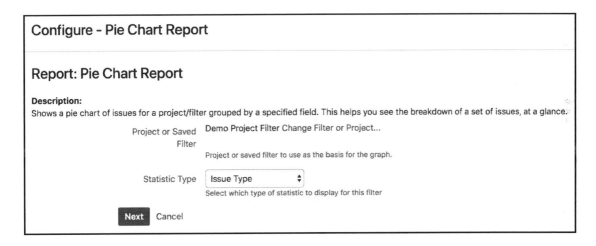

Configure - Pie Chart Report

Report: Pie Chart Report

Description:
Shows a pie chart of issues for a project/filter grouped by a specified field. This helps you see the breakdown of a set of issues, at a glance.

Project or Saved Filter — Demo Project Filter Change Filter or Project...

Project or saved filter to use as the basis for the graph.

Statistic Type — Issue Type ▲▼

Select which type of statistic to display for this filter

Next Cancel

3. Once you have configured the report and clicked on the **Next** button, Jira will generate the report and present it on the screen, as shown in the following diagram:

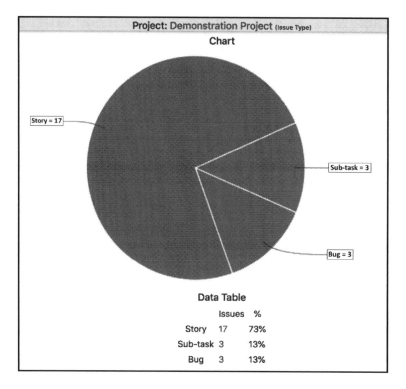

The report type determines the report's layout. Some reports have a chart associated with them (for example, **Pie Chart Report**), whereas other reports will have a tabular layout (for example, **Single Level Group By Report**). Some reports will even have an option for you to export their content into formats such as Microsoft Excel (for example, **Time Tracking Report**).

Dashboards

The dashboard is the first page you see when you access Jira. Dashboards host mini-applications known as **gadgets**, which provide various data and information from your Jira instance. Jira is able to present many of its features, such as filters and reports, on the dashboard using these gadgets, so it is a great way to provide users with a quick one-page view of information that is relevant or of interest to them.

When designing a dashboard, you should always consider the target audience and choose the most appropriate gadgets for the job. For example, a dashboard for the management might have more charts, while a dashboard for a support team can make use of more list style gadgets.

Managing dashboards

When you first install Jira, the default dashboard you see is called the **System Dashboard**, and it is pre-configured to show some useful information, such as all issues that are assigned to you:

1. Since everyone shares the system dashboard, you as a normal user cannot make changes to it, but you can create your own dashboards. Each dashboard's functions are configured independently.
2. Bring down the drop-down menu from **Dashboards**.
3. Select the **Manage Dashboards** option. This will bring you to the **Manage Dashboards** page, as shown in the following screenshot:

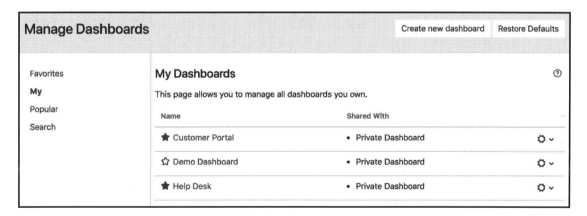

From this page, you can edit and maintain dashboards created by you, search dashboards created and shared by others, and mark them as favorite so that they will be listed as tabs for easy access.

When a dashboard is marked as favorite by clicking on the star icon in front of its name, the dashboard will be accessible when you click on the dashboards link at the top menu bar. If you have more than one favorite dashboard, each will be listed in the tabs and you can select which one to display.

Creating a dashboard

The default **System Dashboard** cannot be changed by normal users, so if you want to have a personalized dashboard displaying information that is specific to you, you will need to create a new dashboard. Perform the following steps to create a new dashboard:

1. Browse to the **Manage Dashboards** page.
2. Click the **Create new dashboard** button.
3. Enter a meaningful name and description for the new dashboard.
4. Select whether you wish to copy from an existing dashboard or start with a blank one. This is similar to creating a new screen from scratch or copying an existing screen.
5. Select whether or not the new dashboard will be a favorite dashboard (for easy access) by clicking on the star icon.
6. Select whether you wish to share the dashboard with other users. If you share your dashboard with everyone by choosing the **Everyone** option, then users that are not logged in will also be able to see your dashboard.
7. Click on the **Add** button to create the dashboard.

The following screenshot shows how you can create a new dashboard from scratch (blank dashboard) and allow view access with all members of the Hummingbird project, and allow edit access to members of **hummingbird-managers** group:

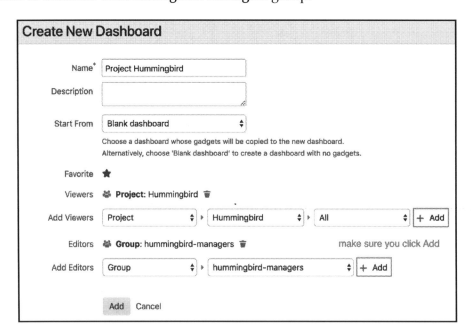

For you to be able to share a dashboard, you will also need to have the **Create Shared Object** global permissions (please refer to Chapter 9, *Securing Jira*, for more information on global permissions).

Once you have created the new dashboard, you will be taken immediately to it. As the owner of the new dashboard, you will be able to edit its layout and add gadgets to it. We will be looking at these configuration options in the next section.

Configuring a dashboard

All custom-created dashboards can be configured once they have been created. As the owner, there are two aspects of a dashboard you can configure:

- **Layout**: This describes how the dashboard page will be divided
- **Contents**: This describes the gadgets that are to be added to the dashboard

Setting a layout for the dashboard

You have to be the owner of the dashboard (created by you) to set the layout. Setting a dashboard's layout is quite simple and straightforward. If you are the owner, you will have the **Edit Layout** option at the top-right corner while you view the dashboard.

Jira comes with five layouts that you can choose from. These layouts differ in how the dashboard page's on-screen real estate is divided. By default, a new dashboard has the second layout that divides it into two columns of equal size:

1. Bring up the drop-down menu from **Dashboards**.
2. Select the dashboard you wish to edit the layout for.
3. Click on the **Edit Layout** option at the top-right corner. This will bring up the **Edit Layout** dialog.

4. Select the layout to which you wish to change, as follows:

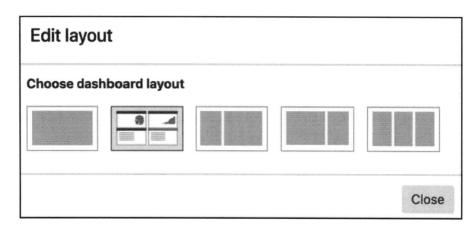

A layout selected from the dialog box will be immediately applied to the dashboard. Any existing contents will automatically have their size and positions adjusted to fit the new layout. After you have decided on your dashboard's layout, you can start adding contents, known as gadgets, onto your dashboard.

Changing the ownership of a dashboard

As with filters, the Jira administrator can change the ownership of a dashboard to a different user, in case the original user has left the organization. Perform the following steps to change a dashboard's ownership:

1. Browse to the Jira administration console.
2. Select the **System** tab and then the **Shared dashboards** option.
3. Search for the dashboard you wish to change ownership of.
4. Click on the **Change Owner** option.
5. Search and select the user that will be the new owner.

6. Click on the **Change Owner** button, as shown in the following screenshot:

Search Shared Dashboards

Search	Owner
Searches in the dashboard's name and description.	Start typing to get a list of possible matches.

Search

Name ◆	Owner	Shared With	Popularity	
Customer Portal	Patrick Li (admin)	• **Group**: jira-software-users (VIEW)	1	⚙ ˅
Help Desk	Patrick Li (admin)	• **Group**: jira-software-users (VIEW)		Change Owner
Project Hummingbird	Patrick Li (admin)	• **Project**: Hummingbird (VIEW) • **Group**: hummingbird-managers (EDIT)		Delete Dashboard

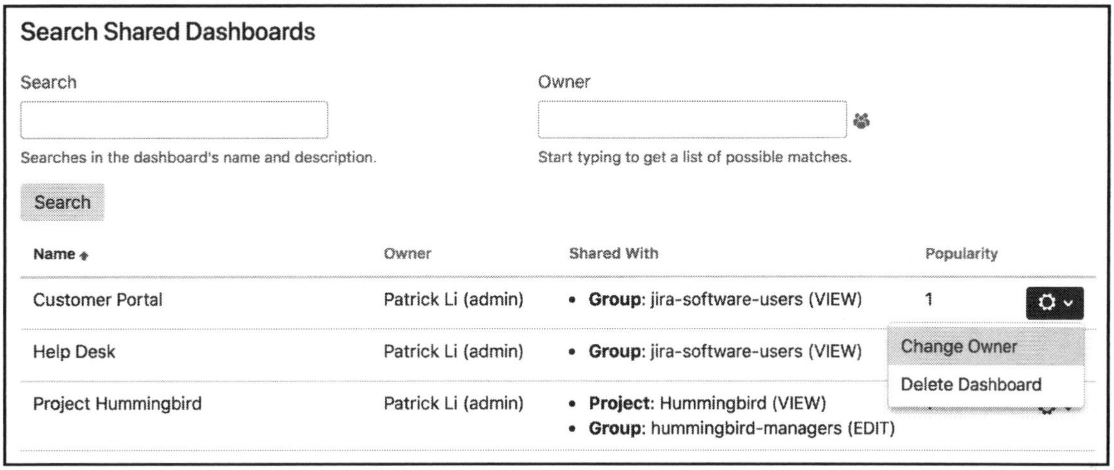

Gadgets

Gadgets are like mini-applications that live on a dashboard in Jira. They are similar to widgets in most of the smartphones we have today or portlets in most portal applications. Each gadget has its own unique interface and behavior. For example, the Pie Chart gadget displays data in a pie chart, while the Assigned to Me gadget lists all the unresolved issues that are assigned to the current user in a table. Gadgets are another way for you to use search filters, by visually presenting the results to end users.

Placing a gadget on the dashboard

All gadgets are listed in the **Gadget Directory**. Jira comes with a number of useful gadgets, such as the **Assigned to Me** gadget that you see on the **System Dashboard**. The following screenshot shows the gadget directory, listing all the bundled gadgets in Jira.

Perform the following steps to place a gadget onto your dashboard:

1. Bring up the drop-down menu from **Dashboards**.
2. Select the dashboard you wish to add a gadget to.
3. Click on the **Add Gadget** option at the top-right corner. This will bring up the **Gadget Directory** window.
4. Click the **Load all gadgets** link to make all gadgets appear.
5. Click on the **Add gadget** button for the gadget you wish to add.
6. Close the dialog to return to the dashboard, as shown in the following screenshot:

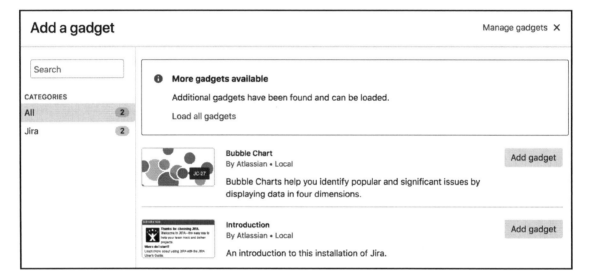

Depending on the gadget you selected, some gadgets will require additional options to be configured. For these gadgets, you will be presented with their configuration screen on the dashboard. Fill in the options and click on the **Save** button.

Let's look at this screenshot of the configuration screen for the **Filter Results** gadget:

On the configuration screen, you can select the search filter to display and control the number of results to show and the fields to include. One common parameter is the **Refresh Interval** option, where you can decide how often the gadget can refresh its content or stay static if you choose never. Whenever you refresh the entire dashboard, all gadgets will load the latest data, but if you stay on the dashboard for an extended period of time, each gadget can automatically refresh its data, so the content will not become stale over time.

Moving a gadget

When you add a gadget, it's usually added to the first available spot on the dashboard. This sometimes might not be where you want to display the gadget on the dashboard, and in other cases, you might want to move the existing gadgets around from time to time. As the owner of the dashboard, you can easily move gadgets on a dashboard through a simple drag-and-drop interface:

1. Browse to the dashboard that has gadgets you wish to move
2. Click on the gadget's title and drag it to the new position on the dashboard

As soon as you drop the gadget to its new location (by releasing your mouse button), the gadget will be moved permanently until you decide to move it again.

Editing a gadget

After configuring your gadget when you first place it on your dashboard, the gadget will remember this and use it to render its content. You can update the configuration details or even its look and feel, as follows:

1. Browse to the dashboard that has gadgets you wish to update.
2. Hover over the gadget and click on the down arrow button at the top-right corner of it. This will bring up the gadget configuration menu.
3. Click on the **Edit** option.
4. This will change the gadget into its configuration mode.
5. Update the configuration options.
6. Click on the **Save** button to apply the changes:

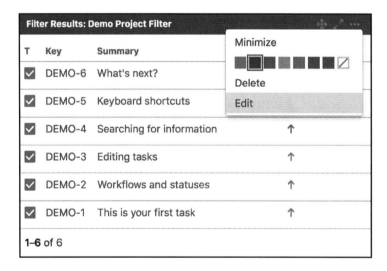

The preceding screenshot shows the **Edit** menu for the **Assigned to Me** gadget. Some gadgets will have a **Refresh** option. Since gadgets retrieve their data asynchronously through AJAX, you can use this option to refresh the gadget itself without refreshing the entire page. The edit, delete, and color options are only available to the owner of the dashboard.

Deleting a gadget

As the owner of the dashboard, you can remove the existing gadgets from the dashboard when they are no longer needed. When you remove a gadget from a dashboard, please note that all the other users who have access to your dashboard will no longer see it:

1. Browse to the dashboard that has gadgets you wish to delete.
2. Hover over the gadget and click on the down arrow button at the top-right-hand corner of it. This will bring up the gadget configuration menu.
3. Click on the **Delete** option.
4. Confirm the removal when prompted.

Once removed, the gadget will disappear from the dashboard. If you choose to re-add the same gadget again at a later stage, you will have to reconfigure it again.

The HR project

In our previous chapters and exercises, we built and customized a Jira project to collect data from users. What we need to do now is process and present this data back to the users. The goal we are trying to achieve in this exercise is to set up a dashboard for our HR team, which will have useful information such as statistics and issue listings, which can help our team members to better organize themselves to provide better services to other departments.

Setting up filters

The first step is to create a useful filter that can be shared with the other members of the team and that also acts as a source of data to feed our gadgets. We will use the advanced search to construct our search:

1. Browse to the **Issue Navigator** page
2. Click on the **Advanced** link to switch to advanced search with JQL
3. Type the `project = HR and issuetype in ("New Employee", Termination) and resolution is empty order by priority` code in the JQL search query
4. Click on the **Search** button to execute the search
5. Click on the **Save as** button to bring up the **Save Filter** dialog

6. Name the filter `Unresolved HR Tasks` and click on the **Submit** button
7. Share the filter with the `hr-team` group setup from Chapter 9, *Securing Jira*, by clicking on the **Details** link next to the **Save as** button

This filter searches for and returns a list of unresolved issues of the `New Employee` and `Termination` type from our `HR` project. The search results are then ordered by their priority so that the users can determine the urgency. As you will see in the later steps, this filter will be used as the source of data for your gadgets to present information on your dashboard.

Setting up dashboards

The next step is to create a new dashboard for your help desk team. What you need is a dashboard specifically for your team so that you can share information easily. For example, you can have the dashboard displayed in a large overhead projector showing all the high priority incidents that need to be addressed:

1. Browse to the **Manage Dashboards** page
2. Click on the **Create new dashboard** button
3. Name the new dashboard `Human Resources`
4. Select a **Blank** dashboard as your base
5. Check the new dashboard as favorite
6. Share the dashboard with the `hr-team` group
7. Click on the **Add** button to create the dashboard

In your example, we will use the two default column layouts for your new dashboard. Alternatively, you are free to experiment with other layouts and find the ones that best suit your needs.

Setting up gadgets

Now that you have set up your portal dashboard page and shared it with the other members of the team, you need to start adding some useful information to it. One example would be to have all the unresolved incidents that are waiting to be processed on the dashboard display. Jira has an **Assigned to Me** gadget, which shows all the issues that are assigned to the currently logged-in user, but what you need is a global list irrespective of the assignee of the incident.

Luckily, Jira also has a **Filter Results** gadget, which displays search results based on a search filter. Since you have already created a filter that returns all the unresolved tasks in your HR project, the combination of both will nicely solve your problem:

1. Browse to the `Human Resources` dashboard you have just created
2. Click on the **Add Gadget** option at the top-right corner
3. Click on the **Add gadget** button for the **Filter Results** gadget
4. Select the **Unresolved HR Tasks** filter you created
5. Add any additional fields you wish to add for the **Columns to display** option
6. Enable **Auto refresh** and set the interval to 15 minutes
7. Click on the **Save** button

This will add a new **Filter Results** gadget to your new dashboard, using your filter as the source of data. The gadget will auto-refresh its contents every 15 minutes, so you will not need to refresh the page all the time. You can add some other gadgets to the dashboard to make it more informative and useful. Some other useful gadgets include the **Activity Stream** and **Assigned to Me** gadgets.

Putting it together

This is pretty much all you have to do to set up and share a dashboard in Jira. After you have added the gadget to it, you will be able to see it in action. The great thing about this is that, since you have shared the dashboard with others on the team, they will be able to see the dashboard too. Members of the team will be able to search for your new dashboard or mark it as a favorite to add it to their list of dashboards.

You do have to keep in mind that, if you are using a filter as a source of data for your gadget, you have to share the filter with other users too; otherwise, they will not be able to see anything from the gadget.

Summary

In this chapter, we covered how users can search and report on the data they have put into Jira, which is an essential component for any information system. Jira provides a robust search facility by offering many different search options to users, including quick, simple, and advanced searches. You can save and name your searches by creating filters that can be rerun at later stages to save you from recreating the same search again.

Jira also allows you to create configurable reports on projects or results brought back from search filters. Information can be shared with others through a dashboard, which acts as a portal for users to quickly have a glance at the data kept in Jira.

In the next chapter, we will look at the other application in the Jira family, Jira Service Desk, which helps to change Jira into a fully functional service desk with powerful features, such as the customer portal and SLA management.

11
Jira Service Desk

Jira was originally designed to be a tool to help developers track software bugs, and, over time, it evolved into a general-purpose, task-tracking tool that can be used by all organizations, thanks to its flexibility and extensibility. For this reason, many organizations started to use Jira as a service desk tool by leveraging its powerful workflow feature, and this has gained tremendous popularity. Recognizing this unique use case and its potential, a new product called Jira Service Desk, from Atlassian, was born. Jira Service Desk is a purpose-built solution that sits on top of the Jira platform, transforming it into a fully-fledged service desk solution with unique capabilities.

In this chapter, you will learn the following topics:

- Installing Jira Service Desk
- Creating and branding a new service desk
- Defining and setting up a **service-level agreement (SLA)**
- Creating custom queues for agents to work from
- Integrating with Confluence to set up a knowledge base

Jira Service Desk

In our previous chapters, we have explored Jira's core features, including workflow, custom fields, and screens. It is not hard to see that you can implement Jira Core or Jira Software as a service desk, but by creating new custom fields, screens, and workflows schemes. While Jira is certainly capable of handling the requirements of a service desk, there are still several things to be desired.

For example, the user interface is often too complicated and confusing for business users to simply create a support ticket. Despite our best efforts, there are still way too many options on the screen, most of which are not useful in a service desk environment. Another example is the lack of ability to set up any sort of SLA to ensure a consistent quality of service.

This is where Jira Service Desk comes in. It addresses all the out-of-the-box shortcomings of Jira by providing a clean, intuitive, and user-friendly interface for both the end customers and support team. It also provides many features that you can expect from a service desk solution. As shown in the following screenshot, Jira Service Desk lets you serve your customers in four easy steps:

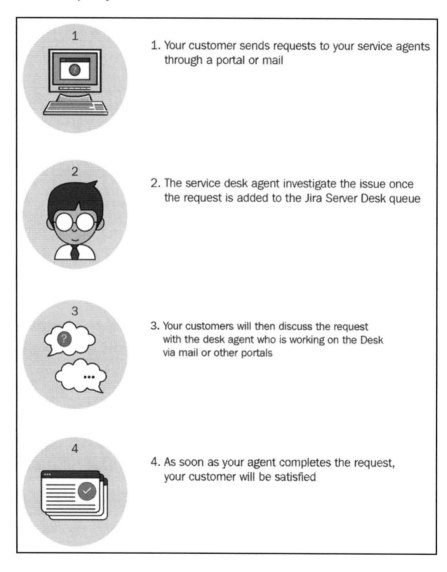

1. Your customer sends requests to your service agents through a portal or mail

2. The service desk agent investigate the issue once the request is added to the Jira Server Desk queue

3. Your customers will then discuss the request with the desk agent who is working on the Desk via mail or other portals

4. As soon as your agent completes the request, your customer will be satisfied

Installing Jira Service Desk

There are two ways you can get Jira Service Desk. The first option is to install it in an existing Jira Core or Jira Software instance that you possess. This is the easiest approach, as it does not require you to provision additional hardware and lets you leverage what you already have. It also makes it easy for your agents to collaborate with other teams to help resolve customer requests. These are the steps you should follow to install Jira Service Desk:

1. Log in as a Jira administrator user.
2. Browse to the Jira administration console.
3. Select the **Applications** tab.
4. Click on the **Try it for free** button under Jira Service Desk from the right-hand panel.
5. Accept the user agreement and follow the onscreen instructions to complete the installation:

The second option is to install Jira Service Desk as a standalone application. Use this option if you do not have a Jira Core or Jira Software instance already running, or if you would like to keep your software issue tracking system and support system separate.

Your agents and other teams can still collaborate to resolve customer requests as in option one, with a few extra steps to set up. These are as follows:

1. Create an application link between the two Jira instances.
2. Integrate both Jira instances with the same user repository, such as LDAP, to ensure that you have the same set of user details in both systems.

To install Jira Service Desk as a standalone application, you can refer to `Chapter 1`, *Getting Started with Jira*, as the installation steps are mostly identical.

Getting started with Jira Service Desk

Before we start using Jira Service Desk, it is important to understand and familiarize ourselves with the key terminology, as follows:

- **Agents**: These are members of your service support team that will be working on customer requests. They are users that can perform actions such as editing, assigning, and closing requests.
- **Customers**: These are the end users that will be raising support requests in your service desk. These can be customers of your product, or colleagues from other departments needing IT support.
- **Customer portal**: This is the main landing page for your customers. It is a simple, clean, and easy-to-use front interface for your service desk, without all the extra noise from the standard Jira interface, as shown in the screenshot below:

Help Center

IT Support

Welcome! You can raise a IT Support request from the options provided.

What do you need help with?   Q

Common Requests

Logins and Accounts

Computers

Applications

Servers and Infrastructure

Get IT help
Get assistance for general IT problems and questions.

Set up VPN to the office
Want to access work stuff from outside? Let us know.

Request a new account
Request a new account for a system.

Desktop/Laptop support
If you are having computer problems, let us know here.

Request a desk phone
If you'd like to request a desk phone, get one here.

Report a system problem
Having trouble with a system?

- **Queues**: These are like Jira filters that show you a subset of issues that meet a certain criterion. Service desk agents use queues to prioritize and pick out requests to work on.
- **Requests**: These are what your end users (not agents), such as customers, submit to Jira Service Desk. Under the hood, they are just normal Jira issues. However, using the term "request" is less confusing in the context of a service desk environment. In short, requests are what your customers see, and issues are what agents see.
- **Service desks**: These are where customers will raise their requests. Under the hood, a service desk is a Jira project of the **Service Desk** project type. Please refer to `Chapter 2`, *Using Jira for Business Projects*, for more information on project types.

As shown in the following screenshot, when customers interact with requests, the user interface is very different to what agents will see. It is much simpler; the UI only displays key information about the request, such as its description and status. Customers cannot make changes to the request details, and can only add new comments or attachments to the request:

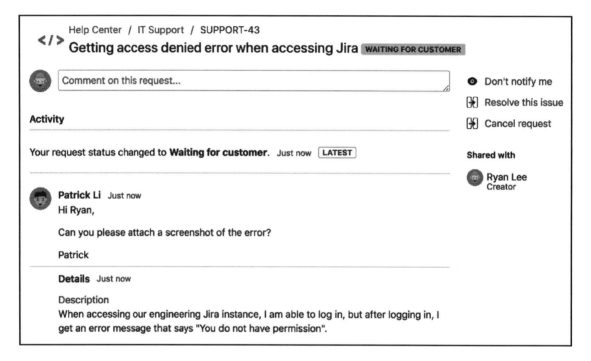

The key information regarding service desks is as follows:

- **Request type**: This represents the different types of request customers can make. These can be anything, including a problem report, help request, or general inquiry. When you create a new request type, Jira creates a new issue type behind the scenes. One major feature of the request type is that it allows you to specify a user-friendly name for it. While the actual issue type is called **Problem Report**, you can rename it and display it as `Submit a problem report` instead.

- **Service desk**: This is what agents will be working from. Each service desk has a front, customer-facing portal. Behind the scenes, a service desk is a Jira project controlled by Jira permissions, workflows, and other schemes.

- **SLA**: SLA defines the quality of service that is being guaranteed to your customers. In Jira Service Desk, SLAs are measured in time, such as response time and overall time taken to resolve issues.

Creating a new service desk

The first step to start working with Jira Service Desk is to create one. Since, under the hood, a service desk is a Jira project with a brand new user interface, you can either create a new service desk from scratch or change an existing project to be of the **Service Desk** project type.

To create a new service desk, perform the following steps:

1. Select the **Create project** option from the **Projects** drop-down menu.
2. Choose a project template, such as **IT Service Desk**, from the **Service Desk** project type, and click on **Next**.
3. Enter the name and key for the new service desk project and click on **Submit**:

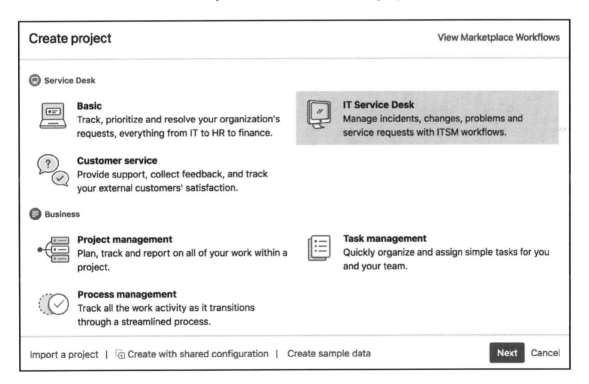

If you choose to use an existing Jira project and convert it into a service desk, all you have to do is update the project's type by following these steps:

1. Browse to the project administration page for the project you want to turn into a service desk.
2. Select the **Change project type** option from the **Actions** menu.
3. Select the **Service Desk** option and click on **Change.**

Once your service desk is created, you will be taken to your service desk user interface, as shown in the following screenshot:

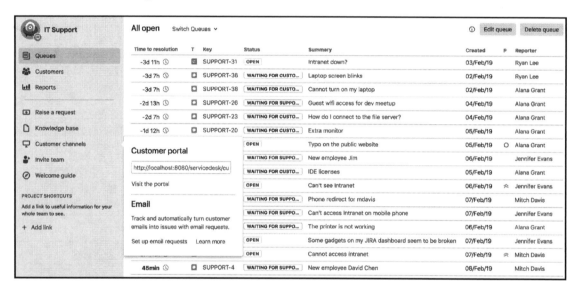

Every service desk has two interfaces. One will be used by you as the admin and members of your support team, called **agents**. The second interface is called the **Customer portal**, which is what customers will see and use to create requests and interact with agents. As you make configuration changes for your service desk, you can always preview the change by clicking on **Customer channels** and then the **Visit the portal** link from the left navigation panel, which will show you what the customer portal will look like.

The URL shown under the **Customer portal** is what your customers should use to access your service desk.

Branding your customer portal

You can brand your customer portal for your service desk with the following options:

- **Help center name**: This is the overall name for your help center. Think of this as the name for your Jira instance.
- **Help center logo**: This is the logo for your help center that will appear in the top-left corner. Think of this as the logo for your Jira. Jira Service Desk will use this logo to automatically change and adjust the top bar color.
- **Customer portal name**: This is the name for a specific service desk portal.
- **Customer portal introduction text**: This is a welcome text that will be displayed for a specific service desk portal.
- **Customer portal logo**: This is the logo for a specific service desk portal.

The following screenshot illustrates each of these items on a sample customer portal:

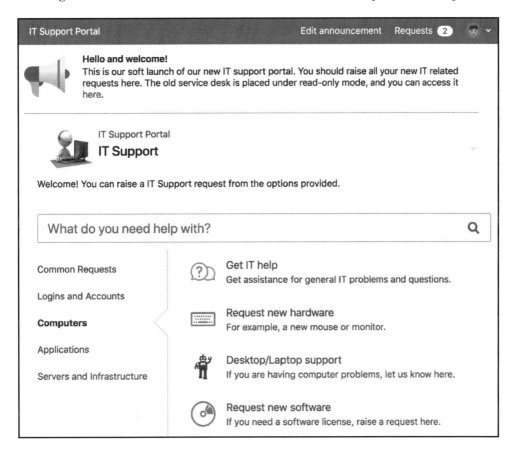

To configure a specific customer portal's branding, perform the following steps:

1. Browse to the project administration page of the service desk you want to brand.
2. Select **Portal settings** from the left-hand panel.
3. Enter a name and welcome text in the **Name** and **Introduction text** fields, respectively.
4. Check the **Use a custom logo for this Customer Portal** option and upload a logo for your customer portal:

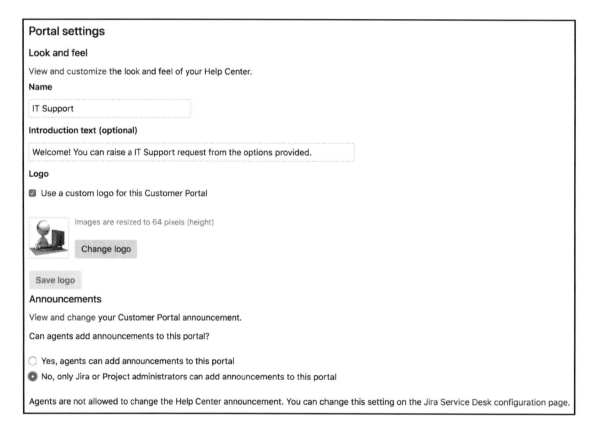

Service desk user types

Jira Service Desk introduces a number of new user types. Under the hood, these user types are mapped to the following new project roles created by the Jira Service Desk when it is installed:

- **Agent**: These are members of the service desk team that work on requests. Agents are added to the **Service Desk Team** project role.
- **Customer**: These are end users that will be submitting requests through your help desk portal. Customers are added to the **Service Desk Customers** project role.
- **Collaborator**: These are the members from other business functions, and are not members of your service desk team, but can help solve customer problems. A good example would be product domain experts or engineers. Collaborators are added to the **Service Desk Team** project role.

Adding an agent to a service desk

Agents are Jira users who will be working on customer requests in Jira Service Desk. These are usually members of your support team. Agents consume the Jira Service Desk licenses. To add an agent to a service desk, do the following:

1. Browse to the service desk you want to add an agent to.
2. Click on the **Invite team** option in the left-hand panel.

3. Search and add users you want to invite as an agent (member) of your service desk team. You can select and add more than one agent. Click on the **Invite** button:

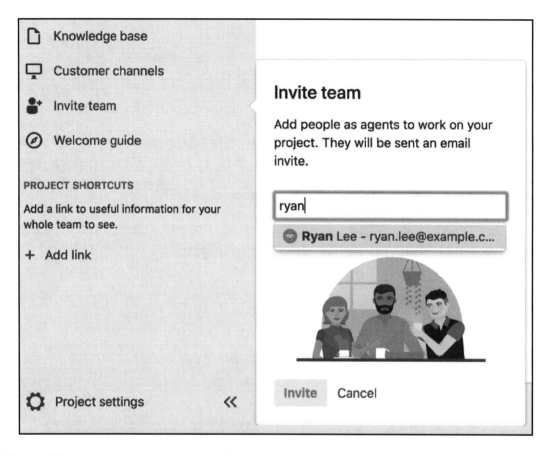

When adding an agent to a service desk, you can select an existing user in Jira, which will grant the user access to the service desk. If the user you want to add as an agent does not exist, you can also create a new Jira account and add them as an agent in a single step by typing in the user's email address. An email will be sent out with a link to set their password. New user accounts created in this way will be automatically added to the **jira-servicedesk-users** group and **Service Desk Team** project role. Refer to Chapter 9, *Securing Jira*, for more information on groups and roles.

Adding a customer to a service desk

Customers are end users who will be creating requests through your customer portal. You can manually invite customers or allow them to sign up themselves. Jira Service Desk requires customers to have an account to submit requests. The good news is that customers do not consume the Jira Service Desk licenses, so you can have as many customers as you want. To invite a customer to a service desk, perform the following steps:

1. Browse to the service desk where you want to add a customer.
2. Select the **Customers** option from the left-hand panel.
3. Click on the **Invite customers** button.
4. Enter the email addresses of customers to invite, and click on the **Send invites** button.

Emails will be sent out to customers with details on how to access the customer portal and steps to create an account if necessary.

If you want to allow users to sign up themselves, you will need to set your service desk to **Everyone can access** and enable the **Anyone can sign up** option, as shown in the following screenshot. If you want to restrict your service desk to just a list of pre-approved customers, then you will need to select the **Only people on my customer list can access my Customer Portal** option:

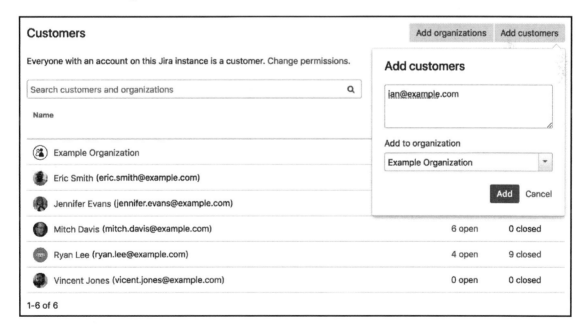

Adding a collaborator to a service desk

Collaborators are Jira users who are not part of your support team (not agents), but have expert knowledge and understanding in the domain area that can assist agents in diagnosing and solving customer requests. In Jira Service Desk, collaborators are users in the **Service Desk Team** project role, but not in the **jira-servicedesk-users** group, and adding a user as a collaborator is an easy way to grant that user access to your service desk project. Collaborators do not consume Jira Service Desk licenses.

To add a collaborator to your service desk, follow these steps:

1. Browse to the project administrator page for the service desk you want to add a collaborator to.
2. Select the **Users and roles** option from the left-hand panel.
3. Click on the **Add users to a role** link.
4. Search and select the users to add, choose the **Service Desk Team** role, and click on the **Add** button.

When making a user a collaborator, you are simply granting permission for the user access to your service desk, so they can view, comment, and add attachments to the request.

Request types

Jira uses issue types to define the purpose of issues, while the Jira Service Desk uses request types for the same purpose. In fact, each request type is mapped to an issue type behind the scenes. The one key difference between the two is that a request type is what is shown to the customers, and a more descriptive display name is used. For example, an issue type is called **Incident**, and the corresponding request type will be called **Report system outage**. You can think of request types as issue types with a more informative display name.

Setting up request types

To create a new request type for your service desk, do the following:

1. Browse to the project administration page for the service desk that you want to create a new request type for.
2. Select the **Request Types** option from the left-hand panel.
3. Select the group this request type belongs to from the left. We will talk about groups later in this section.
4. Click on the icon to select a new icon for the request type.
5. Enter a name for the request type. You can be as descriptive as possible with its name, so your customers can easily understand its purpose.
6. Select the issue type that the request type is mapped to.
7. Enter an optional description. The description will be displayed underneath the request name to help your customer decide what type of request to create.
8. Click on the **Add** button to create the new request type:

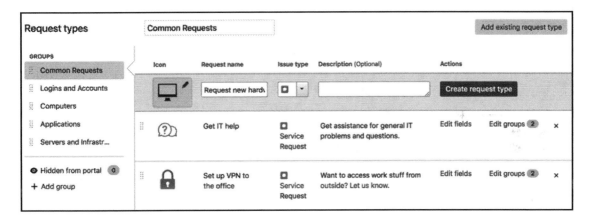

You can reorder the request types by dragging them up and down the list. The order you set in the list will be reflected on the customer portal. Make sure you put some thought into this. For example, you can order them alphabetically or by placing the most common request types at the top.

Organizing request types into groups

As the number of request types grows, you can group similar request types into groups. Therefore, when customers visit the portal, all the request types will be organized logically, making navigation much easier. For example, as shown in the following screenshot of a customer portal, we have six request type groups, where the first five come with Jira Service Desk's project templates, and there's a custom sixth, **Sample Group**. When clicking on **Sample Group**, we have the three request types that customers can raise:

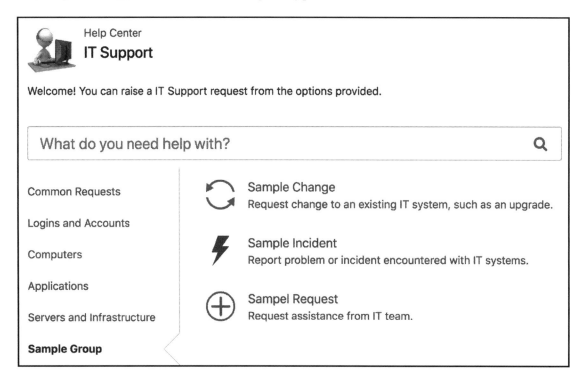

As we have already seen in the *Setting up request types* section, you can add one or more groups to a request type. You can select one of the existing groups, or create a new group, by simply typing in the new group's name. When a request type belongs to two or more groups, it will be displayed when each of the groups is selected on the portal.

Setting up fields

Jira Service Desk lets you set up different field layouts for each request type. The important thing to note here is that, when you are setting up fields for Jira Service Desk, you are not creating new custom fields (as you would in Jira Software); you are simply adding and removing existing fields in the request form when customers create a new request. You can think of this as adding fields onto screens. If you want to add a field that does not yet exist, you will have to create a new custom field first, as described in Chapter 5, *Field Management*, and then make it available in the request form.

Just as with request types, Jira Service Desk allows you to give a custom display name to the field, independent of the actual field's name. This means that the field can be more informative when displayed to customers. For example, for the Jira Summary field, you can give it a display name of What is the problem you are having?. As the display name is independent of the field's name, your existing filters and search queries will continue to work as they are.

To set up field layouts for a request type, follow these steps:

1. Browse to the project administration page for the service desk you want to set up field layouts for.
2. Select the **Request Types** option from the left-hand panel.

3. Click on the **Edit fields** link for the request type you want to set up fields for. This will list all the fields that are currently displayed when customers create a new request:

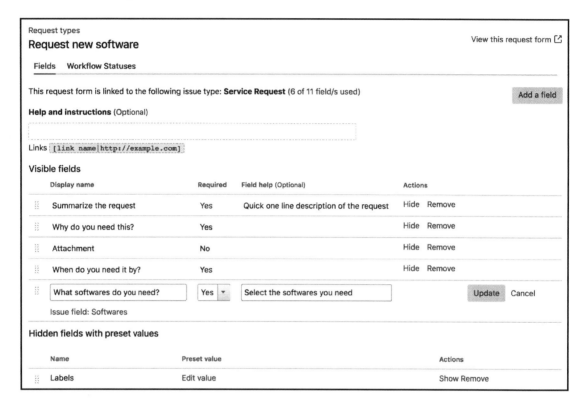

4. Click on the **Add a field** button and select an existing field (both system and custom) to add to the request type.

5. Click on the field's **Display name** to change what customers will see when the field is displayed. This does not change the field's actual name in Jira; it only makes the display more user-friendly.

6. Change the field's mandatory requirement by clicking on the **Required** column. Note that you cannot change this value if it is grayed out, such as the **Summary** field.

After you have set up your field layout for the request type, you can click on the **View this request form link** at the top to see a preview of the result. As shown in the following screenshot, we added the **Due Date** field to the form, but it is now displayed as **When do you need it by?**:

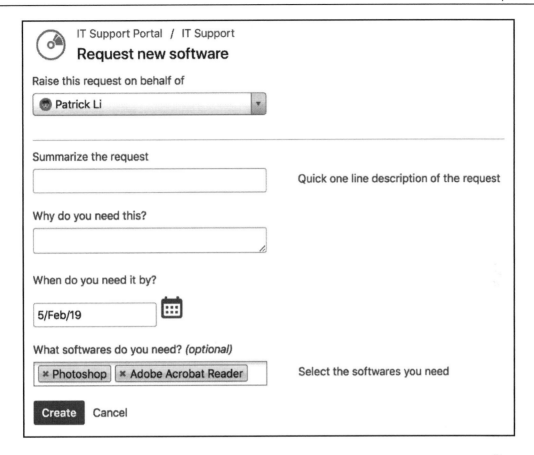

Setting up a workflow in Jira Service Desk

Just as with fields, you can also control how workflow statuses are displayed in Jira Service Desk. Note that you cannot actually change the actual workflow, but you can make the workflow less confusing to your customers, so they know exactly how their requests are progressing.

To set up the workflow for a request type, perform the following steps:

1. Browse to the project administration page for the service desk you want to set up a workflow for.
2. Select the **Request Types** option from the left-hand panel.

3. Click on the **Edit fields** link for the request type you want to set up a workflow for.

4. Select the **Workflow Statuses** tab. This will list all the workflow statuses that are available in the workflow, as shown in the following screenshot:

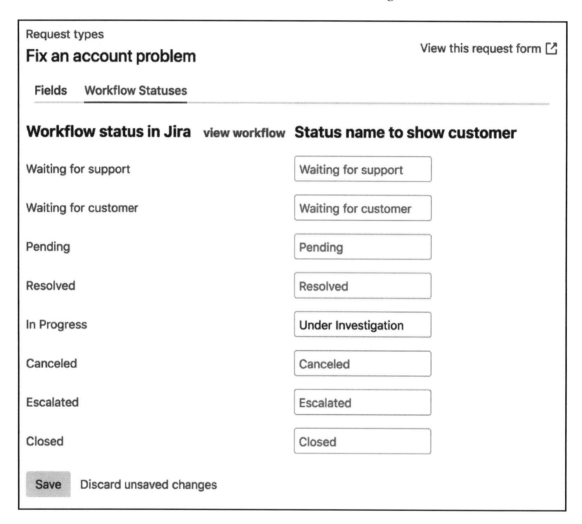

As we can see in the preceding screenshot, the actual Jira workflow status names are listed in the left-hand column. For each of the statuses, you can choose to give it a different display name that will be shown to customers.

For example, the **Open**, **In Progress,** and **Reopened** statuses are normal Jira workflow terms, and represent the fact that the request is currently with a support agent. However, these names can be confusing to customers, so we give them new display names.

 You are not changing the workflow itself. You are simply making it more user-friendly to your customers.

Service level agreement

An SLA defines the agreement between the service provider (your organization) and end user (your customer) in terms of aspects of the service provided, such as its scope, quality, or turnaround time.

In the context of support service, an SLA will define different response times for different types of support requests. For example, severity 1 requests will have a response time of one hour, and severity 2 requests will have a response time of four hours.

Jira Service Desk lets you define SLA requirements based on response time. You can set up the rules on how response time will be measured, and the goals for each rule.

Setting up an SLA

Jira Service Desk's SLA is divided into two components: the time measurement and goals to achieve. Time can be measured for a variety of purposes. Common examples include overall time taken for request resolution and response time to customer requests. To set up an SLA metric, follow these steps:

1. Browse to the project administration page for the service desk you want to set up the SLA on.
2. Select the **SLAs** option from the left-hand panel and then click on the **Create SLA** option.

A simple example will be Jira Service Desk starting to count time as soon as the request is created. Every time an agent requests further information from the customer, the count will be paused until the customer responds. Once the request is finally closed off, the count will be stopped. The following points show you how to set up an SLA time measurement for a simple example:

- For the **Start** column, we will select the **Issue Created** option, indicating that it can start counting time as soon as the request is created
- For the **Pause on** column, we will select the **Status: Waiting for Info** option, indicating that the counting can be paused when the request enters the **Waiting for Info** status
- For the **Stop** column, we will select the **Entered Status: Canceled**, **Entered Status: Closed**, and **Resolution: Set** options, indicating that the counting will be stopped once the request is canceled, closed, or a resolution is set

As you can see in the following screenshot, for each of the three columns, you can select more than one condition:

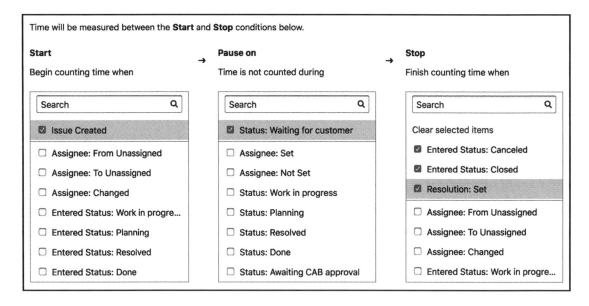

This allows you to set up multiple entry points to start and stop time. An example of this usage will be to measure response time. For example, you will need to guarantee that an agent will respond to a new request within an hour. If the request is sent back to the customer for further information, a response time of one hour is also required as soon as the customer updates the request with the requested information. The following points show you how to set up the time measurement for this SLA:

- For the **Start** column, we will select both the **Issue Create** option and **Entered Status: In Progress** option. Therefore, we will start counting when the issue is first created, and also when it is put back for our agents to work on.
- For the **Stop** column, we will select both the **Entered Status: Waiting for Info** and **Entered Status: Closed** option. Counting will stop when an agent sends the request back to the customer for more information or when it is closed for completion.

The difference between the two examples here is that, in the second example, we do not pause time counting when the request enters the **Waiting for Info** status; instead, we stop counting completely. This means that when the request enters the **Waiting for Info** status, the current counting cycle ends, and when the request enters the **In Progress** status, a new counting cycle will begin, as shown in the following screenshot:

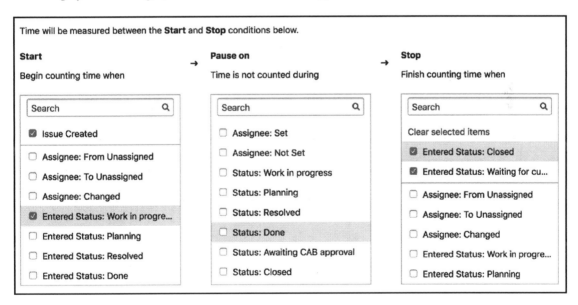

Once we have defined how time should be measured, the next step is to set up the SLA goals. The SLA goals define the amount of time allowed for each of the scenarios we have just set up. If we take the aforementioned response time example, we may set up our goals as shown in the following screenshot:

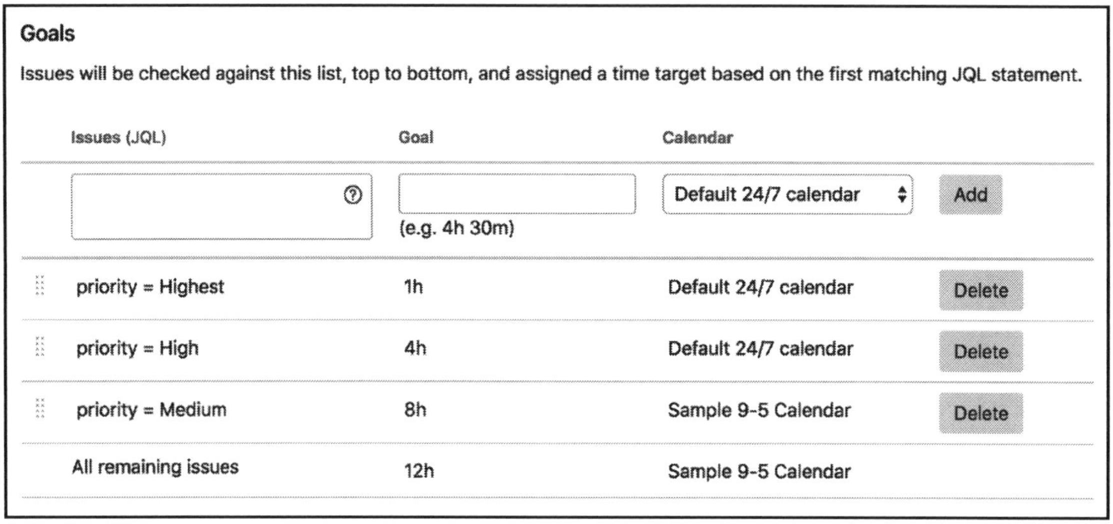

Goals

Issues will be checked against this list, top to bottom, and assigned a time target based on the first matching JQL statement.

Issues (JQL)	Goal	Calendar	
	(e.g. 4h 30m)	Default 24/7 calendar ⬍	Add
priority = Highest	1h	Default 24/7 calendar	Delete
priority = High	4h	Default 24/7 calendar	Delete
priority = Medium	8h	Sample 9-5 Calendar	Delete
All remaining issues	12h	Sample 9-5 Calendar	

In our example, we defined that for requests with priority set to **Highest**, the response time will be 1 hour (**1h**); **High** requests and **Medium** requests will have a response time of 4 and 8 hours, respectively. Everything else will be responded to within 12 hours.

As you can see, there are several components when it comes to defining an SLA goal, which are as follows:

- **Issues**: These are the issues/requests that will have the goal applied to them. Use JQL to narrow down the selection of issues.
- **Goal**: This is the time value for the goal. You can use the standard Jira time notation here, where 3h means 3 hours, 45m means 45 minutes, and 2h30m means 2 hours and 30 minutes.
- **Calendar**: These define the working days and hours the SLA will be applied to. For example, **24/7 Calendar** means that time will be counted every hour of every day. As we will see later, you can create your own custom calendars to define your working day, hours, and even holidays.

When defining SLA criteria, we will need to use JQL. Just like doing an advanced search, Jira Service Desk provides syntax autocomplete to help us validate our queries, as shown in the following screenshot:

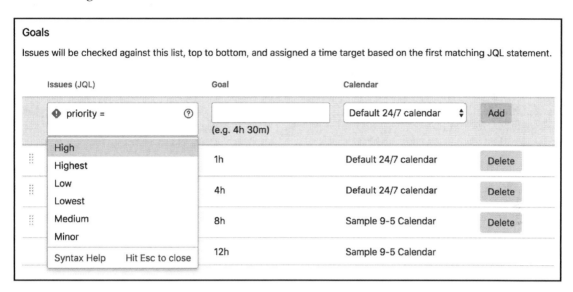

Setting up custom calendars

As we have seen, when setting up an SLA, you can select a calendar that defines the working days and hours, which can be counted toward the goal. Jira Service Desk comes with a **Sample 9-5 Calendar** out of the box, which will only count the time between 9 a.m. and 5 p.m from Monday to Friday.

You can create your own calendars as follows:

1. Browse to the project administration page for the service desk you want to add a calendar for.
2. Select the **SLAs** option from the left-hand panel.
3. Click on the **Calendar** option, and then click on the **Add calendar** button from the dialog box.
4. Enter a name for the new calendar and configure the options.
5. Click the **Save** button to create the calendar.

Jira Service Desk lets you configure your calendar with the following options:

- **Time zone**: This selects the time zone that will be used for the calendar
- **Working days**: This select the days that can be counted toward the SLA
- **Working hours**: These are the hours of each working day that can be included in the SLA
- **Holidays**: This adds holidays, such as Christmas, to be excluded from the SLA

As shown in the following screenshot, we have set up our calendar to have a working time between 9 .a.m. and 5 .p.m, **Tuesday** to **Friday**; this means Monday, Saturday, and Sunday are excluded when calculating SLA metrics. We also added **Christmas Day** and **New Year Day** as holidays, so the SLA will not be applied on those days:

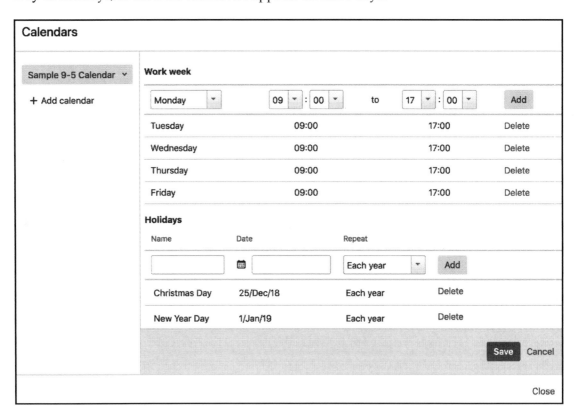

Queues

Queues are lists of requests with predefined criteria for agents to work through. You can think of them as Jira filters. They help you and your teams organize the incoming requests into more manageable groups, so you can better prioritize them. Jira Service Desk uses Jira's search mechanism to configure queues. Refer to `Chapter 10`, *Searching, Reporting, and Analysis*, for more details on Jira search options.

Creating a new queue

You, as the service desk administrator, can create new queues for your team. To create a new queue, follow these steps:

1. Browse to the service desk you want to add a queue for.
2. Select the **Queues** option from the left-hand panel and click the **New queue** option from the **Switch Queues** drop-down menu.
3. Enter a name for the queue. It should clearly reflect its purpose and the types of requests that will be in it.
4. Use the UI controls to create the search criteria. If you are familiar with JQL, or need to use exclusion logics in your query, you can click on the **Advanced** link and use JQL directly.
5. Select the fields that will be displayed when the queue is showing the issue list. Click on the **More** option to find more fields to add. You can also drag the fields left and right to rearrange them. You can select the fields that will display the most useful information.

6. Click on the **Save** button to create the queue, as shown in the following screenshot:

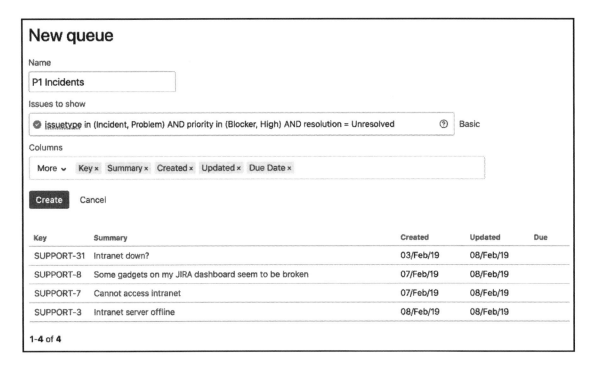

As shown in the preceding screenshot, when you make changes to your search criteria and field selection, there is a preview area at the bottom that will show you the result of your search and the field layout.

Creating knowledge base articles

As your team works diligently to solve problems for your customers, nuggets of knowledge will start to accumulate over time. These include things such as common questions customers face, and the steps taken to troubleshoot them. Jira Service Desk allows you to extract this information and create a knowledge base, which helps customers find solutions themselves. Out of the box, Jira Service Desk only supports Atlassian Confluence for knowledge-base creation, but it is possible to use other tools via third-party add-ons.

To integrate Jira Service Desk with Confluence, you will first have to create an application link between Jira and Confluence. If you have already done this, feel free to skip to the next section. To create an application link for **Confluence**, perform the following steps:

1. Browse to the Jira administration console.
2. Select the **Applications** tab and the **Application links** option from the left-hand panel.
3. Enter the fully-qualified URL to your **Confluence** instance in the **Application** textbox, and click the **Create new link** button, as shown in the following screenshot:

4. Follow the onscreen wizard to complete the linking process.

Once the application link is created with **Confluence**, we can use it for Jira Service Desk. Each service desk will need to be individually integrated with a **Confluence** space. To set up a **Confluence** KB for a service desk, follow these steps:

1. Browse to the project administration page of the service desk you want to set up a **Confluence** knowledge base for.
2. Select the **Knowledge base** option from the left-hand panel.
3. Check the **Link to a Confluence space** option.
4. Select the linked **Confluence** (it may be named something other than **Confluence**) from the **Application** dropdown.
5. Select the Confluence space that the knowledge base article will be created in. If you do not have a space already, click on the **Create a knowledge base space** link.

6. Click on the **Link** button to complete the integration setup, as shown in the following screenshot:

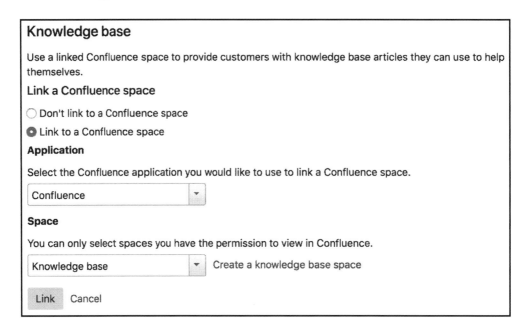

 You can link one service desk to one Confluence space.

After the integration is in place, when an agent views a request, there will be a new **Create KB article** option available. Clicking on that will allow the agent to create a new knowledge base article in the preconfigured Confluence space, as shown in the following screenshot:

From the customer's perspective, a new search box will be available on the customer portal (for a service desk, with the KB feature enabled). Customers will be able to search to see whether there is any information already available in relation to their problems. As shown in the following screenshot, when searching for VPN, the service desk returns a knowledge article from past requests, and if this is what the customer is looking for, it will save valuable time for both the customer and the agent:

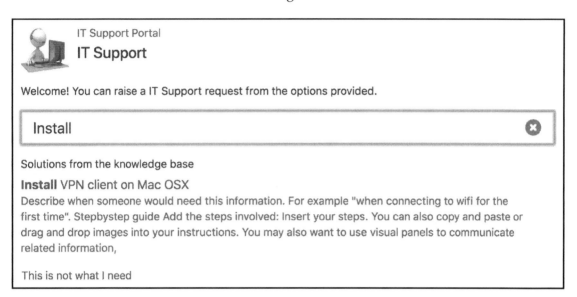

Process automation

When running a service desk, there are many mundane and repetitive tasks that can end up wasting a lot of your team's time. For example, after a request is closed, if the customer subsequently adds a comment, the request needs to be reopened, so it will be placed back into the queue for agents to work on again. Normally, this would require either an agent to manually reopen the request, or you, as the Jira administrator, to configure the workflow used by your service desk project to automatically reopen the request. This can be tedious for the agents, and overwhelming for you, if there are many service desk projects requiring this kind of automation.

The good news is that Jira Service Desk has a process automation feature that greatly reduces the complexity and allows each service desk owner (users with Administer Projects permission) to set up the automation rules, as shown in the following screenshot:

Automation		Add rule
By default rules run as **Patrick Li**. Change default event user		
Name	**Description**	**Actions**
Transition on comment	When a comment is added to an issue, this rule automatically transitions the issue so it's clear who it's waiting on.	View log Edit
Update when a linked issue changes	When the status of an issue changes, this rule will add a comment to its related issues. You can customize this to resolve related issues, change which issues are updated, and more.	View log Edit
Auto-close after being resolved for 3 business days	After 3 business days of the resolution being set, auto-close issues unless the resolution is cleared. The condition and the 3 day limit are set in the 'Time to close after resolution' SLA.	View log Edit
Auto-approve standard changes	After a standard change is created, this rule transitions the change through the 'Peer review / Change manager approval' stage to 'Planning'. It adds a customer-facing comment stating the approval was automated.	View log Edit

Follow the steps shown here to set up automation rules:

1. Browse to the project administration page of the service desk you want to set up automation rules for.
2. Select the **Automation** option from the left-hand panel.
3. Click on the **Add rule** button to create a new automation rule.
4. Select from one of the pre-made automation rule templates from the dialog box, or select the **Custom rule** option from the button to create one from scratch.
5. Enter a name for the new automation rule.
6. Configure the automation rule and click **Save**.

There are a number of things to consider when configuring your automation rule. Firstly, each rule is made up of three parts called **WHEN**, **IF**, and **THEN**, as shown in the following screenshot. The way to think about this is that your rule should read something like this—when something happens on a request, if the criterion is met, then execute the following actions. So, if we take the customer adding comments to a closed request example, the rule may be something like this—when a comment is added, if the request is in the **Closed** status, then transition the request to **Re-opened**.

You configure these components of the automation rule by clicking on the UI elements representing each component. There are a few points to keep in mind when designing your rule:

- You can only have one **WHEN**, which acts as the entry point for the rule. However, it can have multiple triggers, so each rule can be triggered by more than one action.
- You can have more than one **IF** (as **ELSE IF**), so you can set up multiple criteria to evaluate when the rule is triggered.
- You can have only one **THEN**, which can have multiple actions to execute, as shown in the following screenshot:

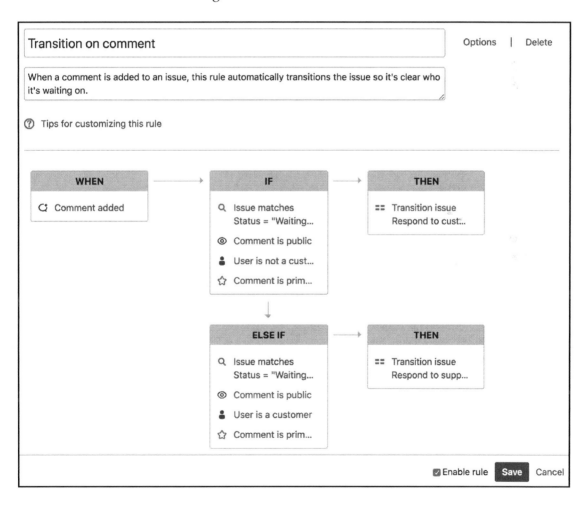

Other options include the following:

- Whether the rule should be run as the user who triggered it, or a dedicated user set for the service desk project. Since not all actions can be run as the user who triggers it, especially if the user is a customer, it is safer to use the project default option.
- Whether the rule can be triggered by another automation rule. This is very useful, as it allows you to chain multiple rules together to automate your process. However, you need to be careful and make sure you do not have rules that will trigger off each other and get stuck in a loop.

Summary

In this chapter, you learned how to use Jira Service Desk to transform Jira into a powerful service desk solution. Jira Service Desk is designed based on many of Jira's out-of-the-box features, such as a workflow engine and search query (JQL), and provides a brand new user interface to remove the friction caused by the old Jira interface. This makes the overall experience a lot more pleasant for customers.

This is the final chapter and the end of our journey. Throughout this book, we looked at the various offerings in the Jira product family and how they can be used to bring value to your organization. Whether you need a project management tool to run agile projects, a service desk solution to support your customers, or simply a tool to better manage and track tasks, Jira has it covered.

We also looked at how you, as an administrator, can install, morph, and adapt it to your environment and use cases. Features such as custom fields and workflows make Jira a very flexible solution that can adapt to your requirements. These features can be further extended by third-party add-ons; we have introduced several popular ones that can bring more capabilities to Jira, and they are just a few out of thousands, more available. Now, it is your job to discover and try other add-ons to enhance your and your users' experience using Jira, and to make it a success.

Other Books You May Enjoy

If you enjoyed this book, you may be interested in these other books by Packt:

Jira Quick Start Guide
Ravi Sagar

ISBN: 9781789342673

- Implement Jira as a project administrator or project manager
- Get familiar with various functionalities of Jira
- Configure projects and boards in your organisation's Jira instance
- Understand how and when to use components and versions in your projects
- Manage project configurations and Jira schemes
- Learn the best practices to manage your Jira instance

Hands-On Agile Software Development with JIRA
David Harned

ISBN: 9781789532135

- Create your first project (and manage existing projects) in JIRA
- Manage your board view and backlogs in JIRA
- Run a Scrum Sprint project in JIRA
- Create reports (including topic-based reports)
- Forecast using versions
- Search for issues with JIRA Query Language (JQL)
- Execute bulk changes to issues
- Create custom filters, dashboards, and widgets
- Create epics, stories, bugs, and tasks

Leave a review - let other readers know what you think

Please share your thoughts on this book with others by leaving a review on the site that you bought it from. If you purchased the book from Amazon, please leave us an honest review on this book's Amazon page. This is vital so that other potential readers can see and use your unbiased opinion to make purchasing decisions, we can understand what our customers think about our products, and our authors can see your feedback on the title that they have worked with Packt to create. It will only take a few minutes of your time, but is valuable to other potential customers, our authors, and Packt. Thank you!

Index

W